Holt California
Physical Science

Standards Review Workbook

HOLT, RINEHART AND WINSTON

A Harcourt Education Company

Orlando • **Austin** • New York • San Diego • London

Printed in the United States of America

ISBN 0-03-093444-3

6 179 09 08

Contents

Introduction

This workbook consists of questions that will help assess students' understanding of the *California Science Standards*. The questions are correlated to the *California Science Standards* for Grade 8 as well as the appropriate *California Science Standards* from Grade 7. This breadth of content coverage provides teachers with an opportunity to assess their students' understanding of the essential science knowledge and skills at the middle school level. These assessments can then help identify topics or concepts in need of re-teaching or additional practice and should be used to inform curricular decisions on the classroom or school levels.

For more practice with the Investigation and Experimentation standards, please refer to the Science Skills Activities in your Student Edition.

California Science
Pre Standards Assessment

1. **The tendency of all objects to resist any change in motion is called**

   A inertia.

   B mass.

   C weight.

   D volume.

2. **Which of the following is not used to describe the motion of an object?**

   A force

   B direction of motion

   C speed

   D position

3. **A hurricane in the Gulf of Mexico travels 360 km in 15 h. What is the average speed of this hurricane?**

   A 30 km/h

   B 24 km/h

   C 18 km/h

   D 12 km/h

4. **Which of the following parts of an atom has no electrical charge?**

   A electron

   B neutron

   C proton

   D the proton and the neutron

5. **Your knee joint is an example of**

   A a sliding joint.

   B a ball-and-socket joint.

   C a hinge joint.

   D a fixed joint.

6. **The graph below shows the distance Kyle traveled on his bicycle trip and the amount of time it took him to travel that distance. During which 5-minute interval did Kyle achieve the greatest average speed?**

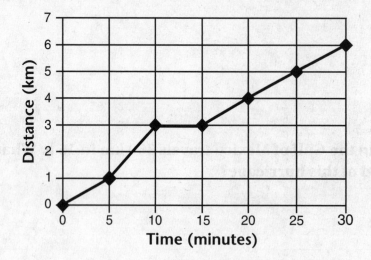

   A 0–5

   B 15–20

   C 10–15

   D 5–10

7. Which of the following is an illustration of the human body using a lever for mechanical advantage?

    A You close your hand into a fist.

    B You kick a soccer ball.

    C You blink your eyes.

    D You wiggle your toes.

8. A teenager pulls a rope to the left with a force of 12 N. A child pulls on the other end of the rope to the right with a force of 7 N. The child's friend adds a force of 8 N, also pulling to the right. What will happen?

    A The net force will be 3 N to the right.

    B The net force will be 15 N to the left.

    C The net force will be 12 N to the right.

    D The net force will be 27 N to the left.

9. When Mendeleev arranged elements in order of increasing atomic mass, he discovered that

    A the properties of the elements repeated in an orderly pattern.

    B all the elements in a group had different properties.

    C there was no pattern in the properties of the elements.

    D all the elements in a period had similar properties.

10. Where are the nonmetals located on the modern periodic table?

   A next to the zigzag line on the table

   B to the right of the metalloids on the table

   C at the left-hand end of the table

   D spread evenly throughout the table

11. Which of the following is an example of a chemical change?

   A water freezing into ice

   B ice melting into water

   C butter softening into a liquid

   D meat spoiling and changing color

12. Which of the following is a characteristic property?

   A density

   B mass

   C volume

   D height

13. A scientist carries out a reaction in a test tube. After the bubbling stops, she notices that the test tube is very warm. What might she conclude about the reaction?

   A The reaction happened very quickly.

   B The reaction is endothermic.

   C The reaction is exothermic.

   D No reaction took place.

14. A transparent object that forms an image by bending light is called a

  A mirror.

  B prism.

  C refractor.

  D lens.

15. The amount of force exerted on a given area is called

  A a pascal.

  B a kilogram.

  C pressure.

  D a newton.

16. What opening lets light into the eye?

  A pupil

  B iris

  C retina

  D rods

17. During a laboratory experiment, Carlos is given a solution that has a pH of 10. Which of the following is a valid conclusion?

  A The solution will turn blue litmus paper red.

  B The solution will taste sour.

  C The solution is a base.

  D The solution will not conduct an electric current.

18. **Which is an example of friction that is helpful?**

    A car engine parts wearing out

    B tires moving a car forward on a road

    C holes developing in your socks

    D the erosion of soil by wind

19. **According to the illustration below, which planet has the oldest surface?**

    Planet A
    115 craters/km²

    Planet B
    75 craters/km²

    Planet C
    121 craters/km²
    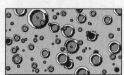

    Planet D
    97 craters/km²
    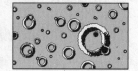

    A Planet A

    B Planet B

    C Planet C

    D Planet D

20. **Gravity is a force of attraction between objects that is due to their masses. Which of the following is true about gravity?**

    A It exists only between Earth and the sun.

    B It exists between all objects in the universe.

    C It affects only objects that touch.

    D It affects only large objects in space.

21. Carbohydrates are biochemicals that are composed of one or more

  A saturated hydrocarbons.

  B sugar molecules.

  C organic compounds.

  D starch molecules.

22. One of the ways scientists classify the stars is according to their color. Which of the following determines the color of a star?

  A its distance from Earth

  B its luminosity

  C its temperature

  D its size

23. Which of the following explains the difference between speed and velocity?

  A One has motion, and the other does not.

  B One has direction, and the other does not.

  C One involves time, and the other does not.

  D One involves acceleration, and the other does not.

24. An object that does not transmit any light is known as

    A translucent

    B transparent

    C opaque

    D reflective

25. In a vacuum, all types of electromagnetic waves have the same

    A speed.

    B pitch.

    C wavelength.

    D amplitude.

26. What occurs when light beams reflect at the same angle?

    A regular reflection

    B irregular reflection

    C diffuse reflection

    D angular reflection

27. **What prevents blood from flowing backward in veins?**

    A platelets

    B valves

    C muscles

    D cartilage

28. **Which of the following is *not* an example of acceleration?**

    A a car turning

    B a bicycle speeding up as it goes down a hill

    C a bus moving from rest to a speed of 40 km/hr

    D a spaceship traveling at 400 km/hr

29. **When two forces exerted on an object are balanced, the object**

    A speeds up.

    B slows down.

    C moves at constant velocity.

    D does not experience a change in motion.

30. A certain substance has a definite shape and volume and its particles do not move fast enough to overcome the attraction between them. What do these properties indicate about the state of the substance?

   A It is a solid.

   B It is a liquid.

   C It is a gas.

   D It is a plasma.

31. The table below shows data from a laboratory experiment in which Anne measured the temperatures of the following substances: ice, ice water, water at room temperature, and boiling water. Which of the following is a valid conclusion from this experiment?

Temperature of Substances

Substance	Temperature
Ice	−2°C
Ice water	0°C
Water	27°C
Boiling water	100°C

   A The particles in ice water have more energy than the particles in ice.

   B The particles in ice have the greatest attraction to each other.

   C The particles in ice water move more quickly than the particles in boiling water.

   D The particles in boiling water have the most energy.

32. Stars in the Milky Way galaxy are light-years apart, but the planets are much closer. For example, it takes light from the sun about 43 minutes to reach Jupiter. What unit best describes distances between planets in our solar system?

A a light-second

B a light-minute

C a light-day

D a light-year

33. A baseball travels a distance of 90 meters in 4.5 seconds. What is the average speed of the baseball in meters per second?

A 9 m/s

B 20 m/s

C 25 m/s

D 45 m/s

34. If the net force on an object is not equal to zero, the forces on the object are

A unbalanced.

B balanced.

C reduced.

D increased.

35. What is the density of a liquid that has a mass of 20 g and a volume of 56 mL?

   A 0.63 g/mL

   B 0.36 g/mL

   C 2.80 g/mL

   D 2.08 g/mL

36. Ions in an ionic compound are arranged in a repeating three-dimensional pattern. What is this pattern called?

   A an ionic solution

   B a chemical bond

   C a valence electron

   D a crystal lattice

37. An atom that has 5 protons, 6 neutrons, and 5 electrons has a mass number of

   A 5

   B 11

   C 10

   D 16

38. Pure water has a pH of

   A 4

   B 10

   C 8

   D 7

39. What must be done for a chemical equation to be balanced?

    A All molecules must be counted.

    B All chemicals must be equal.

    C All atoms must be counted.

    D All atoms must be discounted.

40. When water changes to ice, it is an example of

    A a physical change.

    B a chemical reaction.

    C a chemical bond.

    D a diatomic molecule.

41. Which hydrocarbon contains only single bonds between carbon atoms?

    A saturated

    B aromatic

    C unsaturated

    D inorganic

42. Water pressure increases as depth increases. So, if you place an object in water, the water will exert more pressure on the bottom of an object than it will on the top of the object. The result is a net upward force on the object. What is the name of this net upward force?

A drag

B lift

C buoyant force

D water force

43. The figures below were used to illustrate the arrangement of the particles of a substance in three different states of matter as observed in a laboratory experiment. What physical state of the substance is shown in figure 2?

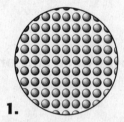 1. 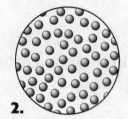 2. 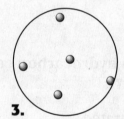 3.

A gas

B liquid

C solid

D vapor

44. Stars are classified by how

    A hot they are.

    B cold they are.

    C far away they are.

    D close they are.

45. Colors that are not transmitted through transparent or translucent objects are

    A absorbed.

    B refracted.

    C reflected.

    D transmitted.

46. Which of the following is true of properties of the elements in the same period of the periodic table?

    A They are similar.

    B They are not similar.

    C They are identical.

    D They have the same atomic mass.

47. What causes a galaxy to have a spiral shape?

    A the stars it contains

    B its rotation

    C gas and dust

    D its mass

48. Which of the following is *not* a primary color of light?

A red

B green

C yellow

D blue

49. What is an organic compound composed only of carbon and hydrogen called?

A molecule

B electron

C hydrocarbon

D single bond

50. Why does a rock sink in water?

A The rock is denser than water.

B Water is denser than the rock.

C Air is denser than the rock.

D Water is less dense than air.

California
Science
Standards
Review

SC7.6.a Students know visible light is a small band within a very broad electromagnetic spectrum.

STANDARD REVIEW

When you look around, you can see things that reflect light to your eyes. It might seem odd to label something that you can't see with the term light. The light that you are most familiar with is called visible light. Visible light is the very narrow range of wavelengths in the electromagnetic spectrum that humans can see. Visible light waves have wavelengths between 400 nm and 700 nm. Visible light energy is changed into chemical energy by green plants during photosynthesis. Ultraviolet light is similar to visible light. Both are forms of energy that travel as a certain kind of wave.

Visible light waves and ultraviolet light waves are both kinds of electro-magnetic waves. Other kinds of EM waves include radio waves, infrared waves, and X rays. Visible light, ultraviolet light, and infrared waves are important to living things. The entire range of electromagnetic waves is called the electromagnetic spectrum. As you can see, visible light is only a small band within the broad electromagnetic spectrum. There is no sharp division between one kind of wave and the next. Some kinds even have overlapping ranges.

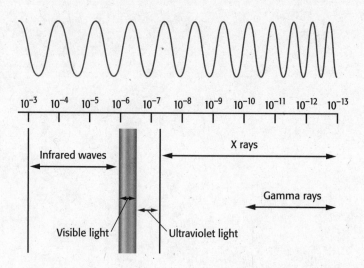

STANDARD PRACTICE

Directions Using the Standard Review and what you have studied, read each question and circle the letter of the best response.

1. **Which of the following statements is true?**

   A Light is a form of electrical energy that travels through wires.

   B Light is a form of energy that travels as a wave and can pass through different media.

   C Light is generated by the sun and is captured for human use.

   D Light is a process by which energy is generated.

2. **Electromagnetic waves that humans can see make up the**

   A visible spectrum.

   B electromagnetic spectrum.

   C sound wave system.

   D longitudinal waves.

3. **What is the entire range of EM waves called?**

   A a rainbow

   B visible light

   C electricity

   D the electromagnetic spectrum

4. **How does light travel?**

   A as matter

   B as a wave

   C as a magnetic field

   D as electric current

SC7.6.b Students know that for an object to be seen, light emitted by or scattered from it must be detected by the eye.

STANDARD REVIEW

If you look at a TV set in a bright room, you see the cabinet around the TV and the image on the screen. But if you look at the same TV in the dark, you see only the image on the screen. The difference is that the screen is a light source, but the cabinet around the TV is not. You can see a light source even in the dark because light emitted by it is detected by your eyes. The tail of the firefly is a light source. Flames, light bulbs, and the sun are also light sources. Objects that emit visible light are called luminous. Most things around you are not light sources. But you can still see them because light from light sources is reflected by the objects and then is detected by your eyes. A visible object that is not a light source is illuminated.

STRUCTURE OF THE EYE

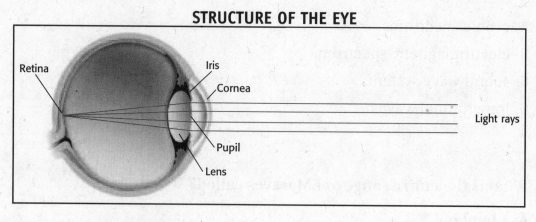

STANDARD PRACTICE

Directions Using the Standard Review and what you have studied, read each question and circle the letter of the best response.

1. **The correct sequence for structures that light passes through when entering the eye, from outside to inside, is**

    A cornea, retina, lens.

    B retina, cornea, pupil.

    C cornea, pupil, lens.

    D pupil, optic nerve, retina.

2. **What determines the color that an object looks to your eyes?**

   A the color of the light

   B the distance of the object from you

   C the wavelengths of light that reach your eyes

   D the amount of light around you

3. **Which light interaction explains why you can see things that do not produce their own light?**

   A absorption

   B reflection

   C refraction

   D scattering

4. **Which of the following is an illuminated object?**

   A the sun

   B a flashlight

   C a firefly

   D the moon

5. **While riding a bicycle, you see a tree branch lying in your path and ride around it. Your eye transmitted electrical impulses to your brain. Your brain then interpreted the information to help you respond. How does your sight help you?**

   A Sight helps you see and avoid dangers.

   B Sight allows you to sense changes in pressure.

   C Sight helps increase the speed of your reflexes.

   D Sight allows you to sense changes in temperature.

SC7.6.c Students know light travels in straight lines if the medium it travels through does not change.

STANDARD REVIEW

Have you ever seen a cat's eyes glow in the dark when light shines on them? Cats have a special layer of cells in the back of their eyes that reflects light. This layer helps the cat see better by giving the eyes a second chance to detect the light. Reflection is one interaction of light waves with matter.

Light travels in straight lines as long as the material that the light travels through doesn't change. So, a ray of light shining through air is usually straight. One way to change the direction of a light beam is by reflection. Reflection is the bouncing back of light rays when they hit an object. But light doesn't change directions randomly. Instead, it follows the law of reflection.

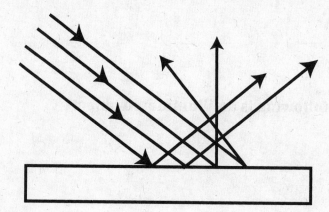

STANDARD PRACTICE

Directions Using the Standard Review and what you have studied, read each question and circle the letter of the best response.

1. **In what direction does your brain see light as traveling when it reflects off an object?**

   A in a straight line

   B in a wavy line

   C in an S-shaped line

   D in a curved line

2. **Why can you see your reflection in a window?**

   A Light is absorbed by the glass.

   B Light is refracted by the glass.

   C Light is reflected off the glass.

   D Light is transmitted through glass.

3. **What happens when light waves bounce off an object?**

   A refraction

   B incidence

   C reflection

   D diffraction

4. **What happens to the colors that are not transmitted through transparent or translucent objects?**

   A The colors are absorbed.

   B The colors are refracted.

   C The colors are reflected.

   D The colors are transmitted.

5. **The change in the direction of a wave when the wave finds an obstacle or an edge, such as an opening, is called**

   A refraction.

   B incidence.

   C reflection.

   D diffraction.

SC7.6.d Students know how simple lenses are used in a magnifying glass, the eye, a camera, a telescope, and a microscope.

STANDARD REVIEW

What do cameras, telescopes, and the human eye have in common? They all use lenses to form images. A lens is a transparent object that forms an image by refracting, or bending, light. Light rays that pass through the center of any lens are not refracted. The point at which beams of light cross after going through a lens is called the focal point. The distance between the lens and the focal point is called a focal length.

A magnifying glass has a convex lens. A magnifying glass can form images of an object that are larger or smaller than the object is. A convex lens, such as the lens in a magnifying glass, forms two kinds of images. The lens of the human eye is also a convex lens. This lens refracts light and focuses the light on the back surface, or retina, of the eye. The muscles that hold the lens of the eye can change the shape of the lens to help it focus images. The cornea of the eye also refracts light.

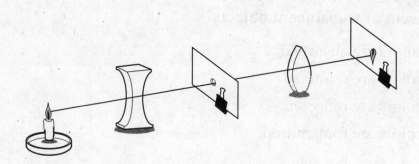

STANDARD PRACTICE

Directions Using the Standard Review and what you have studied, read each
question and circle the letter of the best response.

1. **Which of the following is a true statement about the lens of the eye?**

   A It forms a virtual image.

   B It is thinner in its middle than at its edge.

   C It functions like a convex mirror to reflect light onto the retina.

   D It changes thickness to focus on objects at various distances.

2. The adjustable opening that allows light into a camera is the

   A lens

   B shutter

   C f stop

   D aperture

3. The lens in an eye focuses the image of an object on what at the back of the eye?

   A cornea

   B retina

   C cones

   D optic nerve

4. A convex mirror and a concave lens both cause light rays to

   A slow down.

   B spread apart.

   C gain energy.

   D form blurry images.

5. How does using a magnifying lens change what details you can see?

   A It makes objects appear closer to your eyes.

   B It makes objects appear brighter to your eyes.

   C It makes objects appear smaller than the objects are.

   D It makes objects appear larger than the objects are.

6. What happens when the lens focuses light in front of the retina?

   A nearsightedness

   B farsightedness

   C blindness

   D normal vision

SC7.6.e Students know that white light is a mixture of many wavelengths (colors) and that retinal cells react differently to different wavelengths.

STANDARD REVIEW

Some of the energy that reaches Earth from the sun is visible light. The visible light from the sun is white light. White light is visible light of all wavelengths combined. Light from lamps in your home as well as from the fluorescent bulbs in your school is also white light. Cells in the human eye react differently to different wavelengths of light. As a result, humans see the different wavelengths of visible light as different colors.

The longest wavelengths are seen as red light. The shortest wavelengths are seen as violet light. The range of colors is called the visible spectrum. To help you remember the colors, you can use the name ROY G. BiV. The capital letters in Roy's name represent the first letter of each color of visible light: red, orange, yellow, green, blue, and violet. You can think of i in Roy's last name as standing for the color indigo. Indigo is a dark blue color. Though the colors are given separate names, the visible spectrum is a continuous band of colors.

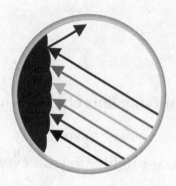

STANDARD PRACTICE

Directions Using the Standard Review and what you have studied, read each question and circle the letter of the best response.

1. **Which of the following is *not* a primary color of light?**

    A red

    B green

    C yellow

    D blue

2. A flashlight beam shines on one end of a prism. A beam containing all the colors of visible light can be seen at the other end. What valid conclusion can be drawn from this observation?

   A The prism refracted the white light, breaking it into the colors of visible light.

   B The prism glass contained all colors, and the light set them free.

   C Light bends as it moves through openings.

   D White light reflects off glass.

3. What property of waves determines the color of a light?

   A amplitude   **C** wavelength

   B mass   **D** wave speed

4. White light is the entire range of

   A colors of visible light combined.

   B ultraviolet light.

   C gamma rays and X rays.

   D EM waves.

5. Which of the following allow you to see the world in color?

   A cones   **C** lenses

   B rods   **D** retina

6. Humans see different wavelengths of light as

   A different colors.   **C** infrared rays.

   B ultraviolet light. **D** matter.

SC7.6.f Students know light can be reflected, refracted, transmitted, and absorbed by matter.

STANDARD REVIEW

When light hits any form of matter, the light can interact with the matter in different ways. The light can be reflected, absorbed, or transmitted. You know that reflection happens when light bounces off an object. Reflected light allows you to see things. And you know that absorption is the transfer of light energy to matter. Absorbed light can make things feel warmer. Transmission is the passing of light through matter. In fact, without the transmission of light, you couldn't see! All of the light that reaches your eyes is transmitted through air and several parts of your eyes. Light can interact with matter in several ways at the same time.

Matter through which visible light is easily transmitted is said to be transparent. Air, glass, and water are examples of transparent matter. You can see objects clearly when you view them through transparent matter. Sometimes, windows in bathrooms are made of frosted glass. If you look through one of these windows, you will see only blurry shapes. You can't see clearly through a frosted window because it is translucent. Translucent matter transmits light but also scatters the light as it passes through the matter. Wax paper is an example of translucent matter. Matter that does not transmit any light is said to be opaque. You cannot see through opaque objects. Metal, wood, and this book are opaque.

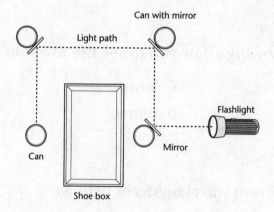

STANDARD PRACTICE

Directions Using the Standard Review and what you have studied, read each question and circle the letter of the best response.

1. **What do you see when sunlight is refracted through water droplets?**

   A a wavelength

   B a rainbow

   C a reflection

   D a light

2. **Whether an object is transparent, translucent, or opaque is determined by its ability to**

   A refract light.

   B diffract light.

   C transmit light.

   D increase the speed of light.

3. **What happens to the speed of light when light travels in matter?**

   A It is slightly faster.

   B It is a lot faster.

   C It is slightly slower.

   D It is a lot slower.

4. **The color of an opaque object is determined by the**

   A the colors of light that are reflected.

   B the colors of light that diffracted.

   C the colors of light that are transmitted.

   D the colors of light that are refracted.

SC7.6.g Students know the angle of reflection of a light beam is equal to the angle of incidence.

STANDARD REVIEW

Light is reflected by surfaces the same way that a ball bounces off the ground. If you throw the ball straight down, it will bounce straight up. If you throw the ball at an angle, it will bounce away at an angle. The law of reflection states that the angle of incidence is equal to the angle of reflection. Incidence is the arrival of a beam of light at a surface.

Why can you see your image in a mirror but not in a wall? The answer has to do with the differences between the two surfaces. A mirror's surface is very smooth. Thus, light beams are reflected by all points of the mirror at the same angle. This kind of reflection is called regular reflection. A wall's surface is slightly rough. Light beams will hit the wall's surface and reflect at many different angles. So, the light scatters as it is reflected. This kind of reflection is called diffuse reflection.

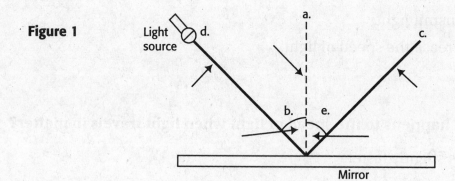

Figure 1

STANDARD PRACTICE

Directions Using the Standard Review and what you have studied, read each question and circle the letter of the best response.

1. **What occurs when light beams reflect at the same angle?**

    A regular reflection

    B irregular reflection

    C diffuse reflection

    D angular reflection

2. The arrival of a beam of light at a surface is called

   A reflection.

   B refraction.

   C incidence.

   D transmission.

3. What interaction of light with matter causes light to change direction?

   A absorption

   B reflection

   C refraction

   D scattering

4. In the diagram below, Fiona has arranged five mirrors that will allow her to watch her friends from behind a wall. Which of her friends is she able to see with this arrangement?

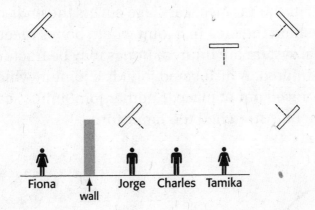

   A Jorge

   B Jorge and Charles

   C Tamika

   D Tamika and Charles

SC7.6.h Students know how to compare joints in the body (wrist, shoulder, thigh) with structures used in machines and simple devices (hinge, ball-and-socket, and sliding joints).

STANDARD REVIEW

A place where two or more bones meet is called a joint. Your joints allow your body to move when your muscles contract. Some joints, such as fixed joints, allow little or no movement. Many of the joints in the skull are fixed joints. Other joints, such as your shoulder, allow a lot of movement. Joints can be classified based on how the bones in a joint move.

Your shoulder is a ball-and-socket joint. Ball-and-socket joints are similar to video-game joysticks. A ball-and-socket joint allows a bone to move up, down, forward, backward, and in a complete circle. A hinge joint allows less movement. The hinge joint that joins your thigh and lower leg allows the knee to bend in one direction. This joint is similar to some door hinges. Gliding joints, also called sliding joints, allow the bones in the wrist and the bones in the foot to glide over one another.

Joints are often placed under a great deal of stress. But joints can withstand a lot of wear and tear because of their structure. Joints are held together by ligaments. Ligaments are strong elastic bands of connective tissue. They connect the bones in a joint. Also, cartilage covers the ends of many bones. Cartilage helps cushion the area in a joint where bones meet. Sometimes, parts of the skeletal system are injured. Bones may be fractured, or broken. Joints can also be injured. A dislocated joint is a joint in which one or more bones have been moved out of place. Another joint injury, called a sprain, happens if a ligament is stretched too far or torn.

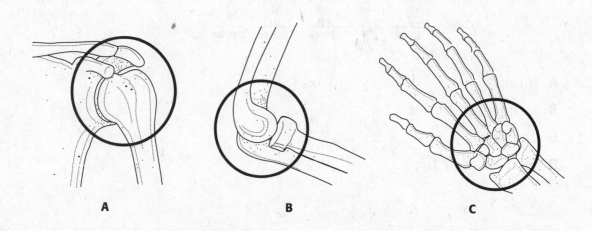

A   B   C

STANDARD PRACTICE

Directions Using the Standard Review and what you have studied, read each
question and circle the letter of the best response.

1. **Which of the joints shown on the left is a ball-and-socket joint?**

   A A

   B B

   C C

   D none of the above

2. **Gliding motion occurs in the joint shown on the left labeled**

   A A

   B B

   C C

   D none of the above

3. **Your knee joint is an example of**

   A a sliding joint.

   B a ball-and-socket joint.

   C a hinge joint.

   D a fixed joint.

4. **What are the three types of joints involved in walking?**

   A the pivot joint, hinge joint, and glide joint

   B the ball-and-socket joint, hinge joint, and glide joint

   C the ball-and-socket joint, hinge joint, and pivot joint

   D the glide joint, hinge joint, and pivot joint

California Science
Standard 7.6.i
Grade 7

SC7.6.i Students know how levers confer mechanical advantage and how the application of this principle applies to the musculoskeletal system.

STANDARD REVIEW

The action of a muscle pulling on a bone often works like a type of simple machine called a lever. A lever is a rigid bar that pivots at a fixed point known as a fulcrum. Any force applied to the lever is called the effort force. A force that resists the motion of the lever, such as the downward force exerted by a weight on the bar, is called the load. In your body, the rigid bar is a bone. The effort force is supplied by muscles. And the fulcrum at which the bone pivots, is a joint. Levers increase the amount of work that can be done by the effort force applied to a load. This increase in work is called mechanical advantage.

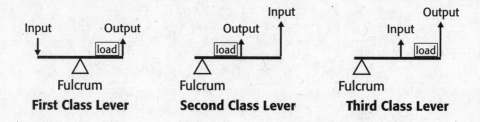

STANDARD PRACTICE

Directions Using the Standard Review and what you have studied, read each question and circle the letter of the best response.

1. **What can you conclude from the location of the fulcrum in the first class lever shown above?**

   A It cannot increase force or distance.

   B It determines where the load and input force are.

   C It is not important.

   D It determines the mechanical advantage.

2. **How does the third class lever shown in the illustration on the left make work easier?**

   A by changing the direction of the force

   B by increasing both force and distance

   C by increasing force and decreasing distance

   D by decreasing force and increasing distance

3. **What would the mechanical advantage be of the third class lever shown in the illustration on the left?**

   A less than 1

   B 1

   C more than 1

   D There is not enough information to determine the answer.

4. **Which of the following items is the same type of lever as the third class lever shown in the illustration on the left?**

   A a seesaw

   B a wheelbarrow

   C a bottle opener

   D an arm lifting a barbell

5. **Which of the following is an illustration of the human body using a lever for mechanical advantage?**

   A You close your hand into a fist.

   B You kick a soccer ball.

   C You blink your eyes.

   D You wriggle your toes.

SC7.6.j Students know that contractions of the heart generate blood pressure and that heart valves prevent backflow of blood in the circulatory system.

STANDARD REVIEW

Your heart, blood, and blood vessels make up your cardiovascular system. Your heart is an organ made mostly of cardiac muscle tissue. When the heart contracts, or squeezes, pressure is created. This pressure moves blood throughout your body.

Flaplike structures called valves are found between the atria and ventricles. Valves are also found where some large blood vessels attach to the heart. As blood moves through the heart, the valves close and produce the "lub-dub, lub-dub" sound of a beating heart. Valves prevent blood from going backward.

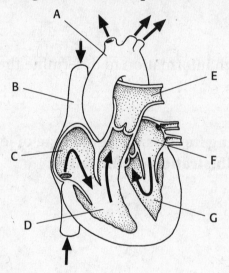

STANDARD PRACTICE

Directions Using the Standard Review and what you have studied, read each question and circle the letter of the best response.

1. **In the illustration above, blood in the chamber labeled *C***

   A is full of oxygen.

   B is returning from the lungs.

   C is oxygen-poor.

   D has very little plasma.

2. The vessel labeled *E* in the illustration on the left, which carries deoxygenated blood, is

   A a pulmonary artery.

   B a pulmonary vein.

   C part of the aorta.

   D part of the atria.

3. The chamber labeled *F* in the illustration on the left is the

   A right atrium.

   B left atrium.

   C right ventricle.

   D left ventricle.

4. As heart ventricles contract and then relax, blood pressure inside of arteries changes from

   A systolic to diastolic.

   B antibody to antigen.

   C type A to type B.

   D plasma to platelet.

5. What prevents blood from flowing backward in veins?

   A platelets

   B valves

   C muscles

   D cartilage

SC8.1.a Students know position is defined in relation to some choice of a standard reference point and a set of reference directions.

STANDARD REVIEW

You know that motion is the movement of an object. But in science, motion has a more precise definition. In science, motion is the change of position of an object relative to a reference point. A reference point is an object that appears to stay in place. To fully describe the motion of an object, you need to describe its position, speed, and direction of the motion.

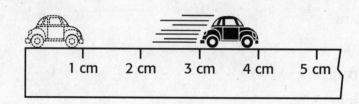

STANDARD PRACTICE

Directions Using the Standard Review and what you have studied, read each question and circle the letter of the best response.

1. **In centimeters per second, what is the speed of the toy car if it moves through the distance shown in 2 seconds?**

    A 0.5 cm/s

    B 1.0 cm/s

    C 1.5 cm/s

    D 2.0 cm/s

    $d = r/t$
    $d = 4/2$

2. **To completely describe the motion of an object, you need**

    A position.

    B speed.

    C direction of motion.

    D all of the above.

3. **You see a dog in a park. Which of the following reference points will best determine if the dog is in motion?**

   A a tree in the park

   B the dog's collar

   C a dog that is running

   D a jogger in the park

4. **If you say "The school is 1 km north of my home," your home is the**

   A relative location.

   B relative position.

   C reference point.

   D reference position.

5. **What is a change in the position of something compared to something that doesn't move?**

   A motion

   B direction

   C reference point

   D relativity

SC8.1.b Students know that average speed is the total distance traveled divided by the total time elapsed and that the speed of an object along the path traveled can vary.

STANDARD REVIEW

To describe the motion of an object, you must determine its speed. Speed is the distance traveled divided by the time taken to travel that distance. Speed is important in describing motion because it tells how fast an object is moving away from its beginning position. Most things do not move with the same speed at all times. For example, runners in a race may have run faster at the end of the race than they did at the beginning. So, it is useful to find the average speed. To calculate average speed, use the equation *average speed=total distance/total time.*

Average Speed	Distance (meters)	Time (seconds)
0	0	0
96	96	1
96	192	2
96	288	3
96	384	4

STANDARD PRACTICE

Directions Using the Standard Review and what you have studied, read each question and circle the letter of the best response.

1. **Why is it useful to find the average speed of an object?**

   A Most things move with the same speed at all times.

   B The speed of most things decreases at a constant rate.

   C Most things do not move with the same speed at all times.

   D The speed of most things increases at a constant rate.

2. **How would you calculate your average speed if you walked from your house to your friend's house two blocks away?**

   A Total distance divided by total time would equal my average speed.

   B Total time divided by total distance would equal my average speed.

   C Total distance times total time would equal my average speed.

   D Total distance plus two divided by total time would equal my average speed.

3. **If it takes ten minutes to walk to your friend's house two blocks away, your average speed is**

   A 0.2 min/block.

   B 0.2 blocks/10 min.

   C 2 blocks/min.

   D 0.2 blocks/min.

4. **The distance traveled divided by the time interval during which the motion occurred is its**

   A motion.

   B reference point.

   C duration.

   D speed.

SC8.1.c Students know how to solve problems involving distance, time, and average speed.

STANDARD REVIEW

The speed of an object is the rate at which the object moves. The speed of an object is rarely constant. So, it is useful to find the average speed of an object. Average speed is the total distance traveled divided by the total time taken. The units for speed are often m/s but can be any distance unit divided by a time unit. The equation below can be used to find average speed.

$$average\ speed = \frac{total\ distance}{total\ time}$$

STANDARD PRACTICE

Directions Using the Standard Review and what you have studied, read each question and circle the letter of the best response.

1. It takes Kira 36 seconds to jog to a store that is 72 meters away. What is her average speed?

   A 1 m/s

   B 4 m/s

   C 2 m/s

   D 0.5 m/s

   $\frac{12}{36} = 2$

2. What is your average speed if you walk 7.5 km in 1.5 hours?

   A 10 km/h

   B 0.2 km/h

   C 5 km/h

   D 8 km/h

   $\frac{7.5}{1.5}$

3. An airplane traveling from San Francisco to Chicago flies 1,260 km in 3.5 hours. What is the airplane's average speed?

A 420 km/h

B 360 km/h

C 500 km/h

D 300 km/h

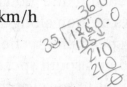

4. An Olympic athlete runs a 400 m race in 50 s. What is her average speed?

A 9 m/s

B 6 m/s

C 8 m/s

D 4 m/s

5. If the average speed of a car is 110 km/h, how much time is needed for the car to travel 715 km?

A 15 h

B 7.15 h

C 5.6 h

D 6.5 h

6. If a horse walked 50 m in 68 s, cantered 150 m in 35 s, and galloped 300 m in 22 s, what would its average speed be?

A 0.25 m/s

B 4 m/s

C 6.22 m/s

D 40 m/s

SC8.1.d Students know the velocity of an object must be described by specifying both the direction and the speed of the object.

STANDARD REVIEW

Velocity can be thought of as the rate at which an object changes its position. The velocity of an object is constant only if the speed and direction of the object do not change. So, constant velocity is always motion along a straight line. The velocity of an object changes if either the speed or direction of the object changes. For example, if a bus traveling at 15 m/s south speeds up to 20 m/s south, its velocity changes. If the bus continues to travel at the same speed but changes direction to travel east, its velocity changes again. And if the bus slows down at the same time that it swerves north to avoid a cat, the velocity of the bus changes yet again.

Speed and direction of motion are combined when describing an object's velocity. Velocity is a quantity that tells both how fast an object is moving (its speed) and which way it is going (its direction of motion). If you know the position of an object and its velocity is constant, you can completely describe the object's motion.

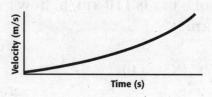

STANDARD PRACTICE

Directions Using the Standard Review and what you have studied, read each question and circle the letter of the best response.

1. **Which of the following is a measurement of velocity?**

   A 16 m east

   B 25 m/s^2

   C 55 m/h south

   D 60 km/h

2. **A bus is moving north at 15 m/s, and you are walking to the rear of the bus at 1 m/s. Your resultant velocity is**

   A 14 m/s north.

   B 16 m/s south.

   C 1 m/s south.

   D 15 m/s north.

3. **A change in a moving object's direction always results in a change in its**

   A motion.

   B speed.

   C force.

   D velocity.

4. **Which of the following does *not* experience a change in velocity?**

   A A motorcyclist driving down a straight street applies the brakes.

   B While maintaining the same speed and direction, an experimental car switches from gasoline to electric power.

   C A baseball player running from first base to second base at 10 m/s comes to a stop in 1.5 seconds.

   D A bus traveling at a constant speed turns a corner.

California Science
Standard 8.1.e
Grade 8

SC8.1.e Students know changes in velocity may be due to changes in speed, direction, or both.

STANDARD REVIEW

Sometimes, the velocity of an object changes. The change in velocity over time is called acceleration. Acceleration can be a change in speed, a change in direction, or both. When a bus driver steps on the gas pedal, the bus will experience acceleration because its speed is increasing. The bus driver can also change the velocity of the bus by turning. In this situation, the bus might not change its speed, but its direction will be different, so it is accelerating. Because acceleration is the change in velocity over time, the units for acceleration look like the units for velocity (m/s) divided by time (s). Thus, the most common units of acceleration are meters per second per second, or (m/s)/s. This unit is often written as m/s^2.

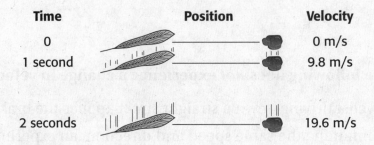

Time	Position	Velocity
0		0 m/s
1 second		9.8 m/s
2 seconds		19.6 m/s

STANDARD PRACTICE

Directions Using the Standard Review and what you have studied, read each question and circle the letter of the best response.

1. **The illustration above shows the results of an experiment that involves dropping a feather and a rock at the same time in a vacuum chamber. Which of the following is true for the rock and the feather?**

    A They are accelerating at the same rate.

    B They are subject to air resistance.

    C They are moving with constant speed.

    D They have the same mass.

2. Given that acceleration is the rate at which velocity changes, what is the rate of acceleration of the rock per second squared?

   A 4.6 m/s

   B 9.8 m/s

   C 19.6 m/s

   D 39.2 m/s

3. Why is an object traveling in a circular motion accelerating?

   A because its direction is always changing

   B because Earth is rotating

   C because nothing else is traveling the same direction

   D because its reference points are also changing

4. A hiker's velocity begins at 1.8 m/s uphill and changes to 1.5 m/s uphill. How do you know that the hiker has a negative acceleration?

   A His direction changed.

   B His direction was unchanged.

   C His speed increased.

   D His speed decreased.

5. To calculate an object's acceleration, you need to know

   A distance traveled and total time.

   B starting point, endpoint, and the object's mass.

   C starting velocity, final velocity, and time it takes to change velocity.

   D average speed and direction traveled.

California Science
Standard 8.1.f

Grade 8

 SC8.1.f Students know how to interpret graphs of position versus time and graphs of speed versus time for motion in a single direction.

STANDARD REVIEW

1. Draw the axes for your graph on a sheet of graph paper. The independent variable is often graphed on the *x*-axis. The dependent variable is often graphed on the *y*-axis.

2. Determine the scale for the axes, and write the numbers of the scale on the axes. Label the axes with appropriate units.

3. Use the data to place data points at the appropriate coordinates on the graph.

4. Use a ruler or a straight edge to draw lines connecting the data points. Write the title of your graph at the top of the graph.

5. Use the graph to identify patterns in the data. Calculating the slope can help you develop quantitative statements about the relationship between variables. Changes in the slope show changes in the relationship between variables.

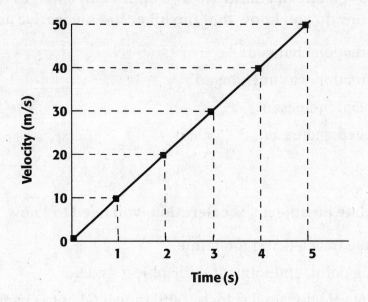

48 Holt California Physical Science, Standards Review Workbook

STANDARD PRACTICE

Directions Using the Standard Review and what you have studied, read each question and circle the letter of the best response.

1. **On the left is a graph of a roller coaster car moving up a hill. What does the straight upward slope show?**

   A positive velocity   **C** positive acceleration

   B negative velocity   **D** negative acceleration

2. **In the sentence "By interpreting the data in the graph, the scientist learned that the speed of the car had been constant," what does the word *interpreting* mean?**

   A figuring out the meaning of   **C** disposing of

   B interacting with   **D** identifying the parts of

3. **On the graph of Your Bug below, the distance from the starting line to Point A is 20 cm. How long did it take your bug to travel that distance?**

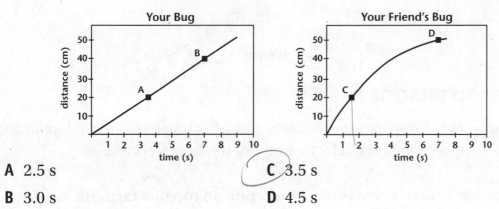

   A 2.5 s   **C** 3.5 s

   B 3.0 s   **D** 4.5 s

4. **Look at the graph of Your Friend's Bug above. How long did it take your friend's bug to travel a similar distance from the starting line?**

   A 1.0 s   **C** 2.0 s

   B 1.5 s   **D** 2.5 s

SC8.2.a Students know a force has both direction and magnitude. A force acts in a particular direction and has a certain size.

STANDARD REVIEW

In science, a force is a push or a pull. All forces have two properties: direction and magnitude. A newton (N) is the unit used to describe the magnitude, or size, of a force. All forces act on objects. For any push to occur, something has to receive the push. The same is true for any pull. Your fingers exert forces to pull open books or to push the keys on a computer. So, the forces act on the books and keys. A force can change the velocity of an object. This change in velocity can be a change in speed, direction, or both. In other words, forces can cause acceleration. Anytime you see a change in an object's motion, you can be sure that the change was caused by a force. However, a force can act on an object without causing the object to move. For example, the force that you exert when you sit on a chair does not cause the chair to move. The chair does not move because the floor exerts a balancing force on the chair.

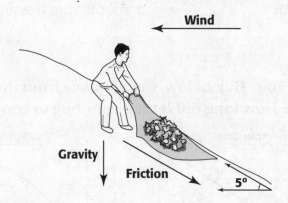

STANDARD PRACTICE

Directions Using the Standard Review and what you have studied, read each question and circle the letter of the best response.

1. **Jorge has raked leaves into a heavy pile on top of a tarp. He wants to pull the tarp up an incline. The sketch below shows the main forces (other than Jorge's) acting on the tarp. What would have to be increased to make it easier for Jorge to pull the tarp up the hill?**

   A the tarp's area

   B the tarp's mass

   C the force of the wind

   D the steepness of the hill

2. **All forces are measured by their direction and their size. Another term for the size of a force is**

   A circumference.

   B magnitude.

   C cumulative force.

   D velocity.

3. **When a soccer ball is kicked, the action and reaction forces do not cancel each other out. Why?**

   A The forces are not equal in size.

   B The forces act on different objects.

   C The forces act at different times.

   D The forces are in different directions.

4. **Which of the following choices best represents force?**

   A a push or a pull always causing motion

   B a push or a pull always causing acceleration

   C a push or a pull acting without an object

   D a push or a pull acting on an object

5. **What does a bulldozer exert on a pile of soil?**

   A motion

   B no force

   C force

   D velocity

SC8.2.b Students know that when an object is subject to two or more forces at once, the result is the cumulative effect of all the forces.

STANDARD REVIEW

When two or more forces act on an object, the result is the combined, or cumulative, effect of the forces. So, the object's motion changes as if only one force acted on the object. That one force is the net force. The net force is the combination of the forces acting on an object. How do you find the net force? The answer depends on the directions of the forces.

Suppose that the music teacher asks you and a friend to move a piano. You pull on one end, and your friend pushes on the other end. The forces that you and your friend exert on the piano act in the same direction. So, the two forces can be added to find the net force. But if the forces are in opposite directions, the net force is found by subtracting the smaller force from the larger one.

STANDARD PRACTICE

Directions Using the Standard Review and what you have studied, read each question and circle the letter of the best response.

1. **Look at the diagram below. What is the net force and direction in which the object will move?**

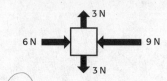

   A 3N to the left

   B 3N to the right

   C 3N downward

   D 3N upward

2. **Look at the diagram below. What is the net force and direction in which the object will move?**

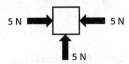

A 5N to the left

B 5N to the right

C 5N downward

D 5N upward

3. **Look at the diagram below. What is the net force and direction in which the object will move?**

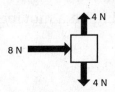

A 8N to the left

B 8N to the right

C 4N downward

D 4N upward

4. **Two forces are acting on an object, but the net force on the object is 0 N. For the net force to be 0 N, all the forces on the object must cancel. What must be true for the two forces on the object to cancel?**

A The forces are the same size and in the same direction.

B The forces are different sizes and in the same direction.

C The forces are the same size and in opposite directions.

D The forces are different sizes and in opposite directions.

SC8.2.c Students know when the forces on an object are balanced, the motion of the object does not change.

STANDARD REVIEW

When the forces on an object produce a net force of 0 N, the forces are balanced. Balanced forces will not cause a change in the motion of an object. Many objects that have balanced forces acting on them are not moving, or are static. Often, tension or compression is acting on these static objects. Tension is a force that is exerted when matter is pulled or stretched. Compression is a force that is exerted when matter is pushed or squeezed. Static objects will not start moving when balanced forces are acting on them. So, a light hanging from the ceiling does not move because the force of gravity pulling down on the light is balanced by the force of tension in the cord pulling upward. An object may also be moving when balanced forces are acting on it. Neither the speed nor the direction of the object will change. A car driven in a straight line at a constant speed is an example of balanced forces acting on a moving object.

STANDARD PRACTICE

Directions Using the Standard Review and what you have studied, read each question and circle the letter of the best response.

1. **If the net forces on an object are balanced**

    A the motion of the object always changes.

    B the motion of the object will not change.

    C the object will always move forward.

    D the object will stop moving.

2. **If the forces on an object are balanced, and the object is standing still**

    A the object will move.

    B the object will remain standing still.

    C the object will increase speed.

    D the object will change direction.

3. **When forces are balanced, the net force equals how many newtons?**

   A 0   C 2

   B 1   D 3

4. **According to Newton's third law of motion, if an ice skater exerts a force on a wall**

   A the wall exerts an equal and opposite force on the skater.

   B the acceleration of the wall depends on the magnitude of the force.

   C the wall will not move because of its inertia.

   D the momentum of the skater is no longer conserved.

5. **How do you know if the forces on an object are balanced?**

   A The object speeds up.

   B The object slows down.

   C The object's motion does not change.

   D The object's direction of motion does not change.

6. **Which of the following statements correctly describes the force that propels the space shuttle upward during takeoff?**

   A The gases push the shuttle upward with a force equal to the shuttle pushing the gases downward.

   B The gases push downward with a force equal to the shuttle pushing the gases downward.

   C The shuttle pushes upward with more force than the gases push downward.

   D The shuttle pushes downward with more force than the gases push upward.

SC8.2.d Students know how to identify separately the two or more forces that are acting on a single static object, including gravity, elastic forces due to tension or compression in matter, and friction.

STANDARD REVIEW

If you push a book so that it slides across a table, you know that the book will eventually stop. What force acts on the book to change its motion? The answer is friction. Friction is a contact force that opposes motion when two surfaces are touching. There are two kinds of friction: static and kinetic.

Gravity is a force of attraction between objects. Weight is a measure of the gravitational force on an object. When you see or hear the word weight, it usually refers to Earth's gravitational force on an object. But weight can also be a measure of the gravitational force exerted on objects by the moon or other planets.

Weight is related to mass, but they are not the same. Weight changes when gravitational force changes. Mass is the amount of matter in an object. An object's mass does not change. Imagine that an object is moved to a place that has a greater gravitational force—such as the planet Jupiter. The object's weight will increase, but its mass will remain the same.

Projectile Motion and Acceleration Due to Gravity

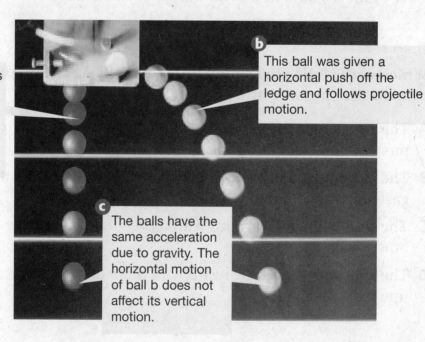

a This ball was dropped with no horizontal push.

b This ball was given a horizontal push off the ledge and follows projectile motion.

c The balls have the same acceleration due to gravity. The horizontal motion of ball b does not affect its vertical motion.

STANDARD PRACTICE

Directions Using the Standard Review and what you have studied, read each
question and circle the letter of the best response.

1. **A measure of the gravitational pull on matter is called**

   A mass.

   B weight.

   C density.

   D volume.

2. **What unbalanced force causes a ball to stop rolling?**

   A friction opposing motion

   B gravity pulling down

   C the ground pushing up

   D the force of the throw or kick

3. **Why does sliding large books take more force than sliding small books?**

   A The large books have more surface roughness.

   B The large books have more weight pushing down.

   C The large books have less air.

   D The large books have less heat.

4. **What two things affect the amount of friction?**

   A energy and heat

   B energy and surface roughness

   C force and heat

   D force and surface roughness

SC8.2.e Students know that when the forces on an object are unbalanced, the object will change its velocity (that is, it will speed up, slow down, or change direction).

STANDARD REVIEW

When the net force on an object is not 0 N, the forces on the object are unbalanced. When unbalanced forces are acting on an object, they cause a change in the velocity of the object. The change in velocity may be a change in speed, a change in direction, or both. Unbalanced forces will also cause a static object to start moving or cause a moving object to slow down and stop moving. Unbalanced forces are needed to change the velocity of moving objects. For example, in a soccer game, the soccer ball is already moving when one player passes it to a second player. When the ball reaches the second player, that player exerts an unbalanced force—a kick—on the ball. After the kick, the ball moves in a new direction and at a new speed.

Unbalanced forces are needed to cause static objects to start moving. Every moving object started moving because an unbalanced force acted on it. But a moving object can keep moving without an unbalanced force acting on it. For example, when it is kicked, a soccer ball receives an unbalanced force. The ball keeps rolling after the force of the kick has ended.

STANDARD PRACTICE

Directions Using the Standard Review and what you have studied, read each question and circle the letter of the best response.

1. **What happens if the forces on an object are unbalanced?**

   A Its motion will not change.

   B Its motion will change.

   C The object will stop moving.

   D The forces will be zero.

2. **When more than one force acts on an object, the forces can reinforce or cancel each other depending on their**

   A direction and size.

   B speed and size.

   C motion and speed.

   D acceleration and velocity.

3. A type of unbalanced force that acts to stop an object in motion is

   A momentum.

   B gravity.

   C inertia.

   D friction.

4. According to Newton's first law, what would happen to a baseball flying through the air if no unbalanced forces acted on it?

   A The baseball would continue in the same direction with the same speed.

   B The baseball would continue with the same speed, but its direction would change.

   C The baseball would continue in the same direction, but its speed would decrease.

   D The baseball would continue in the same direction, but its speed would increase.

5. Which of the following always causes change in speed, direction, or both?

   A balanced forces

   B unbalanced forces

   C either balanced or unbalanced forces

   D any combination of forces

6. An unbalanced force can cause an object's motion to change by

   A changing direction or speed; starting but not stopping motion.

   B changing direction or speed, starting or stopping motion.

   C changing direction or speed; stopping but not starting motion.

   D changing direction or speed only.

SC8.2.f Students know the greater the mass of an object, the more force is needed to achieve the same rate of change in motion.

STANDARD REVIEW

Suppose that you are pushing an empty cart. You have to exert only a small force on the cart to accelerate it. But the same amount of force will not accelerate a full cart as much as it will an empty cart. An object's acceleration decreases as its mass increases and its acceleration increases as its mass decreases.

Mass is a measure of inertia. An object that has a small mass has less inertia than an object that has a large mass. So, changing the motion of an object that has a small mass is easier than changing the motion of an object that has a large mass.

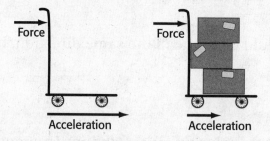

STANDARD PRACTICE

Directions Using the Standard Review and what you have studied, read each question and circle the letter of the best response.

1. **The illustration above shows the experimental setup, and the results observed, for an experiment conducted to measure the acceleration of a wheeled cart. What variable is this experiment designed to test?**

    A amount of mass

    B amount of force

    C size of cart

    D direction of movement

2. **Newton's second law of motion explains why all objects fall with equal acceleration. Which of the following best summarizes this?**

   A Air resistance slows larger objects more than it slows smaller objects.

   B Gravity exerts force on all objects, regardless of their size.

   C The force-to-mass ratio is equal for all objects.

   D Inertia is equal for all objects.

3. **What force is needed to accelerate a 1,500 kg car at a rate of 50 m/s?**

   A 1,550 N   C 1,450 N

   B 75,000 N   D 30 N

4. **Changing the motion of an object with a small mass is easier than changing the motion of an object with a large mass because**

   A an object with a small mass has less inertia than an object with a large mass.

   B an object with a small mass has more inertia than an object with a large mass.

   C smaller objects move more quickly than larger objects.

   D larger objects move more quickly than smaller objects.

5. **A shopping cart will move faster if you give it a hard push than if you give it a soft push because**

   A acceleration decreases as force increases.

   B acceleration increases as force increases.

   C pushing the cart is an action force.

   D pushing the cart is a reaction force.

SC8.2.g Students know the role of gravity in forming and maintaining the shapes of planets, stars, and the solar system.

STANDARD REVIEW

All matter has mass. Gravity is a result of mass. Therefore, all matter is affected by gravity. That is, all objects experience an attraction toward all other objects. This gravitational force pulls objects toward each other. For example, gravity between the objects of the solar system holds the solar system together. Gravity affects smaller objects, too. Right now, because of gravity, you are being pulled toward this book and every other object around you. These objects are also being pulled toward you and toward each other. So, why don't you notice objects moving toward each other? The reason is that the mass of most objects is too small to cause a force large enough to notice. However, you know one object that is massive enough to cause a noticeable attraction—Earth.

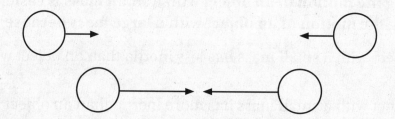

STANDARD PRACTICE

Directions Using the Standard Review and what you have studied, read each question and circle the letter of the best response.

1. **The diagram above was drawn during a study of the relationship between gravitational attraction and distance. Which statement *best* describes the diagram?**

   A As objects are moved closer together, gravitational attraction increases.

   B As objects are moved closer together, gravitational attraction decreases.

   C The distance between objects does not affect gravitational attraction.

   D Gravitational attraction causes objects to move farther apart.

2. One of Newton's laws states that an object traveling at a constant speed in a specific direction will continue to do so unless an unbalanced force acts on it. The moon orbits Earth because

    A no unbalanced force acts on it.

    B an unbalanced gravitational force constantly pulls the moon toward Earth.

    C circular forces act on it.

    D inertia pulls the moon toward Earth.

3. The law of universal gravitation applies to

    A all objects in the universe.

    B only objects on Earth.

    C only the planets that orbit the sun.

    D only certain planets in the solar system.

4. The solar system includes the planets, the moons of those planets, and the sun. Which of the following is true about the solar system?

    A The gravity of the sun does not affect the moons of the planets.

    B Other planets exert gravity on Earth but do not exert gravity on Earth's moon.

    C No gravity exists between Earth and the other planets or their moons.

    D Gravity exists between all parts of the solar system and holds it together.

SC8.3.a Students know the structure of the atom and know it is composed of protons, neutrons, and electrons.

STANDARD REVIEW

Almost all kinds of atoms are made of the same three particles. These particles are protons, neutrons, and electrons. Protons are positively charged particles of the nucleus. Neutrons are the particles of the nucleus that have no electric charge. Neutrons are a little more massive than protons. Protons and neutrons are the most massive particles in an atom. The volume of the nucleus is very small. So, the nucleus is very dense. If it were possible to have a nucleus that has the volume of a grape, that nucleus would have a mass greater than 9 million metric tons!

Electrons are the negatively charged particles in atoms. Electrons are found outside the nucleus in electron clouds. Compared with protons and neutrons, electrons have a very small mass. It takes more than 1,800 electrons to equal the mass of 1 proton. The mass of an electron is so small that the mass is usually thought of as almost zero.

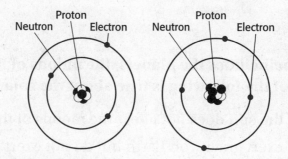

STANDARD PRACTICE

Directions Using the Standard Review and what you have studied, read each question and circle the letter of the best response.

1. **All matter is made up of atoms. Which sentence correctly describes atoms?**

   A All substances are made of the same atoms.

   B Atoms are the smallest particle of a nucleus.

   C An atom is the smallest particle of an element.

   D An atom is a substance that has been cut in half.

2. **The particles inside an atom that are negatively charged are called**

   A protons.

   B neutrons.

   C nuclei.

   D electrons.

3. **What is an atom made up of?**

   A mostly empty space

   B helium

   C gold particles

   D large particles

4. **How would you describe the nucleus?**

   A dense, positively charged

   B large, positively charged

   C tiny, negatively charged

   D dense, negatively charged

5. **Which subatomic particles compose the nucleus of an atom?**

   A electrons and neutrons

   B protons and electrons

   C protons and neutrons

   D protons and ions

SC8.3.b Students know that compounds are formed by combining two or more different elements and that compounds have properties that are different from their constituent elements.

STANDARD REVIEW

What do salt, sugar, baking soda, and water have in common? You might use all of these to bake bread. Is there anything else similar about them? Salt, sugar, baking soda, and water are all compounds. Because most elements take part in chemical changes fairly easily, they are rarely found alone in nature. Instead, they are found combined with other elements as compounds. A compound is a pure substance composed of two or more elements that are chemically combined. Elements combine by reacting, or undergoing a chemical change, with one another. A chemical change, or reaction, happens when one or more substances are changed into one or more new substances that have new and different properties.

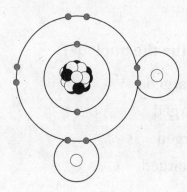

STANDARD PRACTICE

Directions Using the Standard Review and what you have studied, read each question and circle the letter of the best response.

1. **The picture above shows a model of a molecule. Which of the following cannot be determined from the model?**

   A the number of atoms in the molecule

   B the number of electrons in each atom

   C the type of bonds joining the atoms

   D the physical state of the substance

2. **When elements form compounds, the elements**

   A keep their original properties.

   B react to form a new substance with new properties.

   C combine in a random fashion.

   D always change their physical state.

3. **Which pair of atoms can form an ionic bond?**

   A sodium, Na, and potassium, K

   B fluorine, F, and chlorine, Cl

   C potassium, K, and fluorine, F

   D sodium, Na, and neon, Ne

4. **What is true of the new materials formed in a chemical reaction?**

   A Properties differ from original materials.

   B All substances have original properties.

   C Properties are the same as original materials.

   D No substances have original properties.

SC8.3.c Students know atoms and molecules form solids by building up repeating patterns, such as the crystal structure of NaCl or long-chain polymers.

STANDARD REVIEW

The ions that make up an ionic compound are bonded in a repeating three-dimensional pattern called a crystal lattice. In ionic compounds such as table salt, the crystal lattice is built up so that the positive ions are nearest to the negative ions, forming a solid. The shape of the crystals of an ionic compound depends on the pattern of ions in its crystal lattice.

An ionic bond is an attraction between oppositely charged ions. Compounds that have ionic bonds are called ionic compounds. The properties of ionic compounds are a result of strong attractive forces. Ionic compounds can be formed by the reaction of a metal with a nonmetal. Metal atoms become positively charged ions when electrons are transferred from the metal atoms to the nonmetal atoms. This transfer of electrons also causes the nonmetal atom to become a negatively charged ion. Sodium chloride, or table salt, is an ionic compound.

```
          H
          |
        H-C-H
          |
        H-C-H  H
          |    |
        H-C————C-H
          |    |
        H-C-H  H
          |
        H-C-H
          |
          H
```

STANDARD PRACTICE

Directions Using the Standard Review and what you have studied, read each question and circle the letter of the best response.

1. **What kind of carbon backbone does the figure above represent?**

   A a pair chain

   B a ring chain

   C a branched chain

   D a straight chain

2. The ions in an ionic compound are arranged in a repeating three-dimensional pattern called

    A an ionic solution.   **C** a valence electron.

    B a chemical bond.   **D** a crystal lattice.

3. Which of the following is a property of a crystal lattice?

    A low melting point   **C** low boiling point

    B high melting point   **D** malleability

Methane   Ethane   Propane

4. Johanna drew the diagrams shown above of methane, ethane, and propane molecules based on a laboratory experiment. She knows that methane is an organic compound. Which of the following can Johanna conclude holds the atoms in a molecule of methane together?

    A a crystal lattice   **C** covalent bonds

    B acid bonds   **D** ionic bonds

5. Potassium iodide is soluble in water, and has a high melting point of 680°C. Which statement explains how the atoms or ions in this compound are held together?

    A sticky carbon-chained molecules

    B gravitational forces between the nuclei

    C bonds between ions of opposite charges

    D attraction between protons and neutrons

SC8.3.d Students know the states of matter (solid, liquid, gas) depend on molecular motion.

STANDARD REVIEW

You get home from school and decide to make yourself a snack. There are some leftovers in the refrigerator from your dinner last night. So, you heat some up in the microwave oven. As the food heats up, you begin to smell the food. You're also thirsty, so you put some ice in a glass—clink!—and fill the glass with water. You take a big gulp—ahhh!

Each state of matter has a characteristic way in which its particles interact. The scene described above has examples of the three most familiar states of matter: solid, liquid, and gas. The states of matter are the physical forms of a substance and depend on the motion of particles. Matter is made up of very tiny particles called atoms and molecules. Atoms and molecules are in constant motion and are always bumping into each other. The motion of particles is different for each state of matter. The way that the particles interact with each other also helps determine the state of the matter.

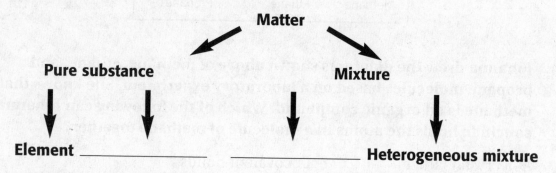

STANDARD PRACTICE

Directions Using the Standard Review and what you have studied, read each question and circle the letter of the best response.

1. **Look at the diagram above. What kind of substance should appear in the blank space under pure substance?**

   A colloid

   B molecule

   C compound

   D mass

2. **Solid, liquid, gas, and plasma are**

   A the four kinds of molecules in matter.

   B the four states of matter.

   C the four elements of matter.

   D the four kinds of atoms in matter.

3. **Which statement about movement in a grain of sand is correct?**

   A Molecules create motion that you can see.

   B The molecules change their positions within the grain.

   C The molecules make movements you can feel but not see.

   D The molecules, in fixed positions, make movements too small to see.

4. **What state of matter has a definite volume but no definite shape?**

   A solid

   B liquid

   C gas

   D plasma

5. **Particles in matter are always moving because they have**

   A elements.

   B states.

   C energy.

   D compounds.

SC8.3.e Students know that in solids the atoms are closely locked in position and can only vibrate; in liquids the atoms and molecules are more loosely connected and can collide with and move past one another; and in gases the atoms and molecules are free to move independently, colliding frequently.

STANDARD REVIEW

A solid is the state of matter that has a definite shape and volume. The particles of a substance in a solid state are very close together. They have a strong attraction between them. The particles in a solid move, but they do not move fast enough to overcome the attraction between them.

Liquid is the state of matter that has a definite volume but takes the shape of its container. The particles in liquids move fast enough to overcome some of the attractions between them. The particles collide with and slide past each other. But the particles remain close together.

Gas is a state of matter that has no definite volume or shape. The particles of a gas have little attraction between them. The particles move about freely and collide randomly with each other. Because gas particles move about freely, the amount of empty space between them can change. The particles of helium in balloons are farther apart than the particles of helium in the tank. As helium particles fill the balloon, they spread apart. The greater amount of empty space between the particles makes the volume of the gas larger.

Jar 1	Jar 2	Jar 3

STANDARD PRACTICE

Directions Using the Standard Review and what you have studied, read each question and circle the letter of the best response.

1. **In the picture above, the circles represent molecules. Which of the jars shows the substance with the most thermal energy?**

   A Jar 1   **C** Jar 3

   B Jar 2   **D** All substances have the same amount of thermal energy.

2. **Enough energy is added to a liquid that the motion of the particles overcomes the attractions between the particles. What is likely to happen to the liquid?**

   A It will change into a solid.

   B It will change into a plasma.

   C It will change into a gas.

   D It will change into a crystal.

3. **Which property of a gas allows you to squeeze an air balloon into smaller shapes?**

   A Gas molecules have a large volume.

   B Gas molecules have a small volume.

   C Gas molecules can be compressed into a smaller volume.

   D Gas molecules collide with one another frequently.

4. **What is the shape of liquids?**

   A square

   B rigid

   C the shape of their containers

   D no shape

SC8.3.f Students know how to use the periodic table to identify elements in simple compounds.

STANDARD REVIEW

Dmitri Mendeleev, a Russian chemist, made a scientific contribution by discovering a pattern to the elements in 1869. Mendeleev saw that when the elements were arranged in order of increasing atomic mass, those that had similar properties fell into a repeating pattern. That is, the pattern was periodic. Periodic means "happening at regular intervals." The days of the week are periodic. They repeat in the same order every seven days. Similarly, Mendeleev found that the elements' properties followed a pattern that repeated every seven elements. His table became known as the periodic table of the elements. The elements are arranged horizontally in order of increasing atomic number. Elements that have similar chemical properties are grouped in vertical columns.

3
Li
Lithium
6.9
11
Na
Sodium
23.0
19
K
Potassium
39.1

STANDARD PRACTICE

Directions Using the Standard Review and what you have studied, read each question and circle the letter of the best response.

1. **Look at the column above, taken from the periodic table. Which of the following statements is correct for the elements shown?**

   A Lithium has the greatest atomic number.

   B Sodium has the least atomic mass.

   C Atomic number decreases as you move down the column.

   D Atomic mass increases as you move down the column.

2. **Which elements often share properties?**

   A those in a period

   B those in a group

   C those with the same color

   D those in a horizontal row

3. **An element is located on the periodic table according to**

   A when it was discovered.

   B its chemical symbol.

   C its chemical name.

   D its physical and chemical properties.

4. **As you move from left to right across the elements in a period in the periodic table, the properties of the elements**

   A are identical.

   B change gradually.

   C become more alike.

   D have greater radioactivity.

5. **At the atomic level, what makes elements reactive?**

   A having a filled outer level

   B exchanging or sharing electrons

   C having the same number of electrons

   D having the same number of protons

SC8.4.a Students know galaxies are clusters of billions of stars and may have different shapes.

STANDARD REVIEW

Large groups of stars, dust, and gas bound together by gravity are called galaxies. Galaxies come in a variety of sizes and shapes. The largest galaxies contain more than a trillion stars. Astronomers don't actually count the stars, of course. Instead, they estimate how many sun-sized stars the galaxy might contain by studying the size and brightness of the galaxy.

About one-third of all galaxies are simply massive blobs of stars. Many look like spheres. Others are more stretched out. These galaxies are called elliptical galaxies. Elliptical galaxies usually have very bright centers and very little dust and gas. Elliptical galaxies contain mostly old stars. Because there is so little freeflowing gas in an elliptical galaxy, few new stars form.

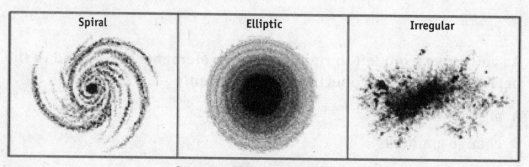

| Spiral | Elliptic | Irregular |

Common Shapes Of Galaxies

STANDARD PRACTICE

Directions Using the Standard Review and what you have studied, read each question and circle the letter of the best response.

1. **Elliptical galaxies and the halos of spiral galaxies contain groups of stars called**

   A globular clusters.

   B open clusters.

   C nebulas.

   D neutron stars.

2. **Which of the following is a piece of evidence that supports the big bang theory?**

   A Most galaxies in the universe are moving away from each other.

   B Most galaxies in the universe are moving away toward the Milky Way Galaxy.

   C Most galaxies in the universe are moving away toward each other.

   D Most galaxies in the universe are moving away toward the sun.

3. **Which of the following is *not* contained in a galaxy?**

   A a globular cluster

   B an open cluster

   C a gas cloud

   D the universe

4. **Why do scientists study distant galaxies to learn about early galaxies?**

   A Distant galaxies are just beginning to form, so they are very similar to early galaxies.

   B Distant galaxies have not changed as much as close galaxies, so they are the most similar to early galaxies.

   C Distant galaxies share many characteristics with early galaxies.

   D Because it takes a long time for light to travel through space, looking at distant galaxies shows what early galaxies looked like.

California Science
Standard 8.4.b

Grade 8

 SC8.4.b Students know that the Sun is one of many stars in the Milky Way galaxy and that stars may differ in size, temperature, and color.

STANDARD REVIEW

The sun is located at the center of the solar system. The sun is 1,390,000 km in diameter and makes up more than 99% of the mass in the solar system. Like other stars in the universe, the sun is made mostly of hydrogen and helium. However, the sun contains traces of almost every other element.

Stars are giant bodies of hot gases that are held together by gravity. Unlike matter on Earth, stars produce their own light. Most stars make light when they combine, or fuse, four atoms of one element, hydrogen, and make another element, helium. This process releases huge amounts of thermal energy and light. If you look closely, you will see that some of the brightest stars show hints of color. Color is one of a star's most basic properties that can be measured. Color helps astronomers classify stars by temperature.

Types of Stars		
Class	Color	Surface temperature (°C)
B	Blue-white	10,000–30,000
F	Yellow-white	6000–7500
G	Yellow	5000–6000
K	Orange	3500–5000

STANDARD PRACTICE

Directions Using the Standard Review and what you have studied, read each question and circle the letter of the best response.

1. **According to the chart above, which color of star is the hottest?**

   A yellow

   B orange

   C yellow-white

   D blue-white

2. **Which of the following is *not* a characteristic of stars?**

    A Stars are made up of hot, dense gas.

    B Stars are all the same size.

    C Stars have apparent motion and actual motion.

    D Stars can be different colors.

3. **Look at the drawing below. Arrange the following according to the way they appear in the life cycle of a star.**

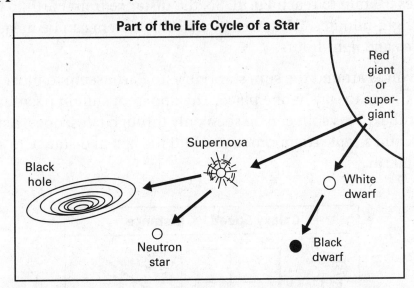

    A supergiant, black hole, supernova

    B supergiant, neutron star, supernova

    C supergiant, black dwarf, white dwarf

    D red giant, supernova, black hole

4. **The sun, like all other stars in our galaxy, changes as it passes through different stages in its life cycle. What kind of star will the sun be in the last stage of its life?**

    A a white dwarf

    B a main-sequence star

    C a red giant

    D a pulsar

SC8.4.c Students know how to use astronomical units and light years as measures of distances between the Sun, stars, and Earth.

STANDARD REVIEW

One way that scientists measure distances in space is by using the astronomical unit. One astronomical unit (AU) is the average distance between the sun and Earth, or approximately 150,000,000 km. Another way to measure distances in space is by using the speed of light. Light travels at about 300,000 km/s in space. Thus, in 1 s, light travels 300,000 km. In 1 min, light travels nearly 18,000,000 km. This distance is called a light-minute. Light from the sun takes 8.3 min to reach Earth. So, the distance from Earth to the sun, or 1 AU, is 8.3 light-minutes. Distances in the solar system can be measured in light-minutes and light-hours.

As Earth revolves around the sun, stars close to Earth seem to move and distant stars seem to stay in one place. The apparent shift in position of a star is called parallax. This shift can be seen only through telescopes. Astronomers use parallax and simple trigonometry to find the actual distance to stars that are close to Earth.

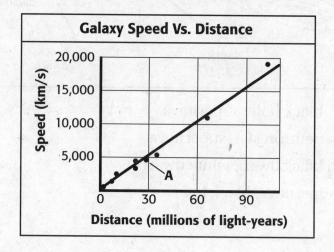

California Science
Standard 8.4.c

Grade 8

STANDARD PRACTICE

Directions Using the Standard Review and what you have studied, read each question and circle the letter of the best response.

1. The graph on the left shows Hubble's law, which relates how far galaxies are from Earth and how fast they are moving away from Earth. How far is galaxy A from Earth?

   A about 5,000 km

   B about 30 million light-years

   C about 10,000 km

   D about 40 million light-years

2. According to the graph on the left, if a galaxy is 90 million light-years from Earth, how fast is it moving away from Earth?

   A 5,000 km/s   C 15,000 km/s

   B 10,000 km/s   D 20,000 km/s

3. Which of the following is used to measure the distance to objects in space?

   A parallax   C zenith

   B magnitude   D altitude

4. The average distance between Earth and the sun is

   A the light-year.   C the kilometer.

   B the astronomical unit.   D the parsec.

Holt California Physical Science, Standards Review Workbook **81**

SC8.4.d Students know that stars are the source of light for all bright objects in outer space and that the Moon and planets shine by reflected sunlight, not by their own light.

STANDARD REVIEW

Stars are giant bodies of hot gases that are held together by gravity. Unlike matter on Earth, stars produce their own light. Most stars make light when they combine, or fuse, four atoms of one element, hydrogen, and make another element, helium. This process releases huge amounts of thermal energy and light. Humans took thousands of years to figure out this very basic secret of why stars shine.

Unlike the sun, the moon does not generate energy that can be emitted as light. Like the planets and other bodies in the solar system, the moon shines because it reflects light from the sun. The total amount of sunlight that the moon gets always remains the same. Half of the moon is always in sunlight, just as half of Earth is always in sunlight. But the moon's period of rotation is the same as its period of revolution. Therefore, you always see the same side of the moon from Earth.

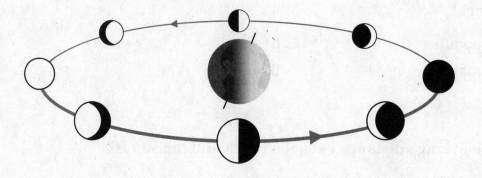

STANDARD PRACTICE

Directions Using the Standard Review and what you have studied, read each question and circle the letter of the best response.

1. **As the moon revolves around Earth, the amount of sunlight that the moon reflects toward Earth changes. When the sun and moon form a 90° angle with Earth, the moon is in which phase?**

   A waning crescent

   B new moon

   C last quarter

   D waxing gibbous

2. **Annular solar eclipses occur because**

   A the moon reflects more sunlight than Earth does.

   B Earth's shadow is too small to completely cover the moon.

   C the moon is too far away from Earth to completely hide the sun.

   D the moon, sun, and Earth are not lined up in a straight line.

3. **A star's magnitude refers to its**

   A temperature.   **C** size.

   B brightness.   **D** age.

4. **How can scientists identify a star's elements?**

   A by its color   **C** by its light

   B by its shape   **D** by its age

SC8.4.e Students know the appearance, general composition, relative position and size, and motion of objects in the solar system, including planets, planetary satellites, comets, and asteroids.

STANDARD REVIEW

The vast region of space called the solar system includes planets, moons, comets, asteroids, meteoroids, and our very own star, the sun. Although Earth may seem large to us, it is only the fifth-largest planet in the solar system. The region of space called the inner solar system contains the four planets closest to the sun: Mercury, Venus, Earth, and Mars. The vast region of space called the outer solar system includes Jupiter, Saturn, Uranus, and Neptune.

Loosely packed bodies of ice, rock, and cosmic dust orbiting the sun are called comets. Scientists think most comets come from the far outer regions of the solar system in a spherical region called the Oort cloud. Between the orbits of Mars and Jupiter lies the asteroid belt. This region of space is about 300 million kilometers across.

Data for the Outer Planets			
Planet	Diameter (km)	Average Distance From the Sun (km)	Period of Revolution (Earth Years)
Jupiter	142,984	778,600,000	12
Saturn	120,536	1,433,500,000	29
Uranus	51,118	2,872,500,000	84
Neptune	49,528	4,495,100,000	164

STANDARD PRACTICE

Directions Using the Standard Review and what you have studied, read each question and circle the letter of the best response.

1. **Most comets seen from Earth follow an orbit that**

    A stays between Mars and Jupiter.

    B is very close to the Earth's orbit.

    C is very elliptical.

    D is nearly circular.

2. Look at the data compiled in the table on the left as part of a field investigation on the solar system. Which of the following is *not* a valid conclusion?

    A In general, each planet is twice as far from the sun as the next closest planet.

    B The period of revolution increases as the average distance from the sun increases.

    C In general, the diameter of the planets decreases as the average distance from the sun increases.

    D In general, the diameter of the planets decreases as the period of revolution increases.

3. Our solar system

    A is not made up of smaller planetary systems.

    B includes the planets, but not the sun.

    C is made up of the inner solar system, the outer solar system, and other smaller systems.

    D is the only known solar system in the universe.

4. Most asteroids in the solar system orbit the sun in the asteroid belt between Mars and Jupiter. Where is the asteroid belt?

    A behind Pluto's orbit

    B in front of Mercury's orbit

    C between the orbits of Neptune and Pluto

    D between the inner planets and the outer planets

SC8.5.a Students know reactant atoms and molecules interact to form products with different chemical properties.

STANDARD REVIEW

A chemical change happens when one or more substances change into new substances that have new and different properties. Chemical changes and chemical properties are not the same. The chemical properties of a substance describe which chemical changes can happen and which chemical changes cannot happen to that substance. But chemical changes are processes by which substances change into new substances. You can learn about a substance's chemical properties by observing which chemical changes that substance can undergo. You see chemical changes more often than you may think. For example, a chemical change happens every time a battery is used. Chemical changes also take place within your body when the food you eat is digested.

A fun way to see what happens during chemical changes is to bake a cake. You combine eggs, flour, sugar, and other ingredients. When you bake the batter, you end up with a substance that is very different from the batter. The heat of the oven and the interaction of the ingredients cause a chemical change. The result is a cake that has properties that differ from the properties of the raw ingredients.

REACTIVITY SERIES OF SELECTED ELEMENTS

Element	Reactivity
K Ca Na	react with cold water and acids to replace hydrogen; react with oxygen to form oxides
Mg Al Zn Fe	react with steam (but not with cold water) and acids to replace hydrogen; react with oxygen to form oxides
Ni Pb	do not react with water; react with acids to replace hydrogen; react with oxygen to form oxides
H_2 Cu	react with oxygen to form oxides
Ag Au	fairly unreactive; form oxides only indirectly

STANDARD PRACTICE

Directions Using the Standard Review and what you have studied, read each
question and circle the letter of the best response.

1. **Based on the reactivity series on the left, which of the following
    elements could create a fire hazard when it is exposed to water?**

    A aluminum

    B sodium

    C silver

    D zinc

2. **In forming rust, an iron nail is reactive with**

    A rubbing alcohol.

    B other iron nails.

    C wood in a house.

    D oxygen in the air.

3. **What happens to the substances involved in a chemical change?**

    A They keep their identities.

    B They change in form.

    C New substances are formed.

    D The substances combine and mix.

4. **Which of the following is *not* the result of a chemical change?**

    A soured milk

    B rusted metal

    C ground flour

    D digested food

SC8.5.b Students know the idea of atoms explains the conservation of matter: In chemical reactions the number of atoms stays the same no matter how they are arranged, so their total mass stays the same.

STANDARD REVIEW

Atoms are never lost or gained in a chemical reaction. They are just rearranged. Every atom in the reactants becomes part of the products. When writing a chemical equation, make sure that the total number of atoms of each element in the reactants equals the total number of atoms of that element in the products. This process is called balancing the equation. Balancing equations comes from the work of a French chemist, Antoine Lavoisier. In the 1700s, Lavoisier found that the total mass of the reactants was always the same as the total mass of the products. Lavoisier's work led to the law of conservation of mass. This law states that mass is neither created nor destroyed in chemical and physical changes. This law means that the total mass of the reactants is the same as the total mass of the products. So, a chemical equation must show the same numbers and kinds of atoms on both sides of the arrow even though the atoms are rearranged.

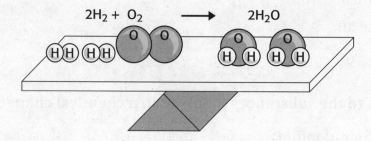

STANDARD PRACTICE

Directions Using the Standard Review and what you have studied, read each question and circle the letter of the best response.

1. **How many atoms are represented in the formula $CaCO_3$?**

   A three

   B four

   C five

   D six

2. **Which of the following is true for the number of atoms involved in the reaction shown in the drawing on the left?**

   A There must be more atoms before the reaction than after the reaction.

   B There must be more atoms after the reaction than before the reaction.

   C The number of atoms at the start of the reaction must equal the number of atoms at the end of the reaction.

   D Atoms at the beginning of the reaction must be used up and not exist at the end of the reaction.

3. **Which of the following states the law of conservation of mass?**

   A Atoms are rearranged in a chemical reaction, and some join new molecules.

   B Two compounds combine to form a new compound with different properties.

   C Mass cannot be created or destroyed in a chemical reaction.

   D Energy is neither created nor destroyed in a chemical reaction.

4. **Which describes what happens to atoms in a chemical reaction?**

   A sometimes lost, never rearranged

   B sometimes lost, gained, or rearranged

   C never lost or gained, just rearranged

   D lost or gained, never rearranged

SC8.5.c Students know chemical reactions usually liberate heat or absorb heat.

STANDARD REVIEW

Chemical energy is part of all chemical reactions. Energy is needed to break chemical bonds in the starting substances. As new bonds form in the final substances, energy is released. By comparing the chemical energy of the original substances with the chemical energy of the final substances, you can decide if energy is released or absorbed in the overall reaction.

A chemical reaction in which energy is released is called an exothermic reaction. Exothermic reactions can give off energy in several forms. If heat is released in an exothermic reaction, the nearby matter will become warmer. The nearby matter absorbs the heat released by the reaction. A chemical reaction in which energy is taken in is called an endothermic reaction. The energy taken in during an endothermic reaction is absorbed from the surroundings.

1 methane + oxygen → carbon dioxide + water + energy
$$CH_4 + 2O_2 \rightarrow CO_2 + 2H_2O + energy$$

2 nitrogen + oxygen + energy → nitrogen oxide
$$N_2 + O_2 + energy \rightarrow 2NO$$

STANDARD PRACTICE

Directions Using the Standard Review and what you have studied, read each question and circle the letter of the best response.

1. **Look at the two equations above. Which of the following statements is true?**

   A Equation 1 is an endothermic reaction.

   B Equation 2 is an endothermic reaction.

   C Both equations 1 and 2 are exothermic reactions.

   D It is not possible to determine the type of reaction.

2. **What happens in an endothermic reaction?**

   A energy is destroyed

   B energy is released

   C energy is created

   D energy is taken in

3. **Which of the following shows an exothermic reaction?**

   A $2Na + Cl_2 \rightarrow 2NaCl + energy$

   B $2H_2 + O_2 \rightarrow 2H_2O$

   C $2H_2O + energy \rightarrow 2H_2 + O_2$

   D $6CO_2 + 6H_2O + energy \rightarrow C_6H_{12}O_6 + 6O_2$

4. **Which states the law of conservation of energy?**

   A In a chemical reaction, energy is neither created nor destroyed, but can change in form.

   B Energy is needed to break chemical bonds.

   C Light energy is released by some exothermic reactions.

   D An exothermic reaction releases energy.

SC8.5.d Students know physical processes include freezing and boiling, in which a material changes form with no chemical reaction.

STANDARD REVIEW

Boiling is the change of a liquid to a vapor, or gas, throughout the liquid. The temperature at which this change happens is the boiling point of the liquid. The change of state from a liquid to a solid is called freezing. Evaporation is the change of state from a liquid to a gas. Evaporation can happen at the surface of a liquid. For example, when you sweat, your body is cooled through evaporation. Condensation is the change of state from a gas to a liquid. Condensation and evaporation are the reverse of each other. The condensation point of a substance is the temperature at which the gas becomes a liquid. The condensation point is the same temperature as the boiling point at a given pressure.

Sublimation is the change of state in which a solid changes directly to a gas. For a solid to change directly to a gas, the particles of the substance must go from being very tightly packed to being spread far apart. So, the attractions between the particles must be completely overcome. The substance must gain energy for the particles to overcome their attractions.

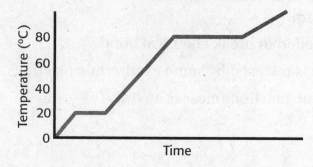

STANDARD PRACTICE

Directions Using the Standard Review and what you have studied, read each question and circle the letter of the best response.

1. **Which of the following is *not* an example of a chemical reaction?**

   A milk turning sour

   B food being digested

   C a match burning

   D ice melting

2. **What occurs when water freezes?**

   A a physical change

   B a chemical reaction

   C a chemical bond

   D a diatomic molecule

3. **In a laboratory investigation on changes of state of matter, Melissa observes that the melting point of water is 0°Celsius and that the freezing point of water is 0°Celsius. What can Melissa conclude from this observation?**

   A Both melting and freezing are exothermic reactions.

   B Water boils at 0°Celsius.

   C Melting and freezing can occur at the same temperature.

   D She made an error in her measurements.

4. **What is happening in the graph below at the point where the line is horizontal?**

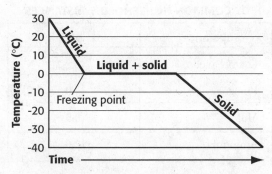

   A Particles have acquired enough energy to slide past one another.

   B Particles are rearranging from a solid into a liquid.

   C Particles are rearranging from a liquid into a solid.

   D Particles are no longer locked in position.

SC8.5.e Students know how to determine whether a solution is acidic, basic, or neutral.

STANDARD REVIEW

You can use the characteristics of a solution to determine if it is acidic or basic. Lemons contain a substance called an acid. One property of acids is a sour taste. A sour taste is not the only property of an acid. Have you noticed that when you squeeze lemon juice into tea, the color of the tea becomes lighter? This change happens because acids cause some substances to change color. An acid is any compound that increases the number of hydronium ions, H_3O^+, when dissolved in water. Hydronium ions form when a hydrogen ion, H^+, separates from the acid and bonds with a water molecule, H_2O, to form a hydronium ion, H_3O^+.

An indicator commonly used in the lab is litmus to identify whether a solution contains an acid or a base. To describe how acidic or basic a solution is, scientists use the pH scale. The pH of a solution is a measure of the hydronium ion concentration in the solution. A solution that has a pH of 7 is neutral. A neutral solution is neither acidic nor basic. Paper strips containing litmus are available in both blue and red. When an acid is added to blue litmus paper, the color of the litmus changes to red.

pH OF COMMON HOUSEHOLD SUBSTANCES

Substance	pH
Lemon juice	2.2
Vinegar	4.0
Milk	6.5
Ammonia	12.0

STANDARD PRACTICE

Directions Using the Standard Review and what you have studied, read each
question and circle the letter of the best response.

1. **What is any compound that increases the number of hydronium
    (H_3O^+) ions dissolved in water called?**

    A base   **C** indicator

    B acid   **D** neutral

2. **Two commonly used indicators for identifying solutions are
    bromthymol blue and**

    A hydrochloric acid.   **C** litmus paper.

    B silver nitrate.   **D** color changer.

3. **What is the reaction between acids and bases called?**

    A neutralization reaction   **C** explosive reaction

    B strength reaction   **D** evaporating reaction

4. **When an acid and a base neutralize each other, what remains?**

    A a weak acid and base   **C** water and a salt

    B a strong acid and base **D** a lipid and a protein

5. **What color does blue litmus paper turn when acid is added to it?**

    A green   **C** blue

    B red   **D** orange

SC8.6.a Students know that carbon, because of its ability to combine in many ways with itself and other elements, has a central role in the chemistry of living organisms.

STANDARD REVIEW

Carbon atoms combine in many ways with each other and atoms of other elements and form the backbone of many molecules needed by living things. Would you believe that more than 90% of all compounds belong to a single group of compounds? This group is made up of covalent compounds that are based on the element carbon. All living organisms need these compounds. For example, the carbon compound glucose, a kind of sugar, is an energy source for many living things.

Carbon has a central role in the chemistry of living organisms. You can understand why when you look at the way carbon makes chemical bonds. Carbon atoms can form long chains with other carbon atoms. Some compounds have hundreds or thousands of carbon atoms! Carbon atoms can also bond with atoms of other elements. Each carbon atom has four valence electrons. So, each carbon atom can make a total of four bonds. Carbon-based molecules can come in many different shapes. Many organic compounds are based on the kinds of carbon backbones shown below. These models show how atoms are connected. Each line represents a covalent bond.

Three Kinds of Carbon Backbones

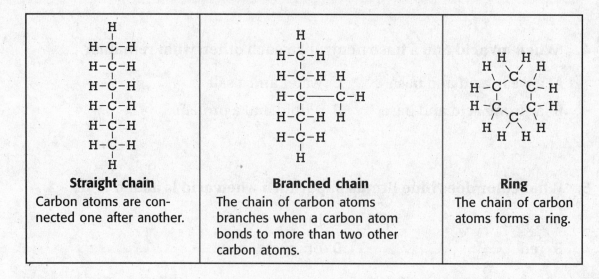

Straight chain	Branched chain	Ring
Carbon atoms are connected one after another.	The chain of carbon atoms branches when a carbon atom bonds to more than two other carbon atoms.	The chain of carbon atoms forms a ring.

STANDARD PRACTICE

Directions Using the Standard Review and what you have studied, read each question and circle the letter of the best response.

1. **How many valence electrons does each carbon atom have?**

   A three

   B two

   C six

   D four

2. **The backbones of some compounds have hundreds or thousands of**

   A carbon atoms.

   B carbon molecules.

   C structural formulas.

   D acid ions.

3. **What type of hydrocarbon has double and triple bonds between carbon atoms?**

   A saturated

   B aromatic

   C unsaturated

   D neutral

4. **Which of the following make up over 90% of all known compounds?**

   A ionic compounds

   B organic compounds

   C basic compounds

   D aromatic compounds

SC8.6.b Students know that living organisms are made of molecules consisting largely of carbon, hydrogen, nitrogen, oxygen, phosphorus, and sulfur.

STANDARD REVIEW

All living things depend on organic compounds. A single cell in your body contains a very large number of organic compounds. Other elements combine with carbon to make the large variety of compounds on which living things depend. Although billions of compounds make up your body, just a few elements make up most of those compounds.

Carbon can combine with elements other than itself and hydrogen. These include oxygen, nitrogen, sulfur, and phosphorus. Your body is made up of a lot of oxygen. Water has atoms of oxygen, and there is a lot of water in your body. Oxygen is found in many organic molecules. You probably don't think much about elements such as sulfur and phosphorus in your body. But they're there! Atoms of nitrogen and sulfur are important parts of proteins. And without phosphorus, your cells could not get the energy that they need. All of these other elements in your body come from the food you eat.

STANDARD PRACTICE

Directions Using the Standard Review and what you have studied, read each question and circle the letter of the best response.

1. **What element is found in the highest percentage by weight in the human body?**

    A hydrogen

    B oxygen

    C carbon

    D phosphorus

2. **Why does carbon have a central role in the chemistry of living organisms?**

   A Carbon helps organisms to regulate temperature.

   B Carbon atoms allow organisms to transmit nerve impulses.

   C Carbon is the most abundant element in organisms.

   D Carbon atoms combine in many ways with atoms of carbon and other elements.

3. **In photosynthesis, what two things do plants use with sunlight to make food?**

   A water and oxygen

   B water and sugar

   C water and carbon dioxide

   D water and salt

4. **In cellular respiration, what do cells use to release energy from food?**

   A water

   B sunlight

   C oxygen

   D carbon dioxide

5. **What element is the most widely used compound in the chemical industry?**

   A sulfur

   B selenium

   C tellurium

   D polonium

SC8.6.c Students know that living organisms have many different kinds of molecules, including small ones, such as water and salt, and very large ones, such as carbohydrates, fats, proteins, and DNA.

STANDARD REVIEW

Biochemicals are very important for the functions of living organisms. In addition to very large molecules, such as carbohydrates, fats, proteins, and DNA, organisms also need a variety of much smaller compounds that are not organic compounds. The most important of these is one that you are very familiar with—water! No living thing on Earth can survive without water. In fact, about 70% of your body is made up of water!

If you have ever tasted a drop of sweat, then you know that salt is present in your body. And if you lose salt through sweating, you need to replace it. But why is salt so important? Salt plays an important role in nerve cells. Salt helps nerve cells conduct electrical signals throughout your body. Ordinary table salt is the kind of salt that your body uses the most. Other salts may also be found in food.

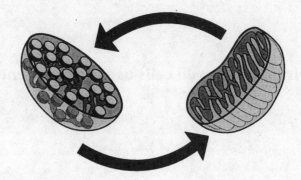

STANDARD PRACTICE

Directions Using the Standard Review and what you have studied, read each question and circle the letter of the best response.

1. **Which of the following compounds is *not* released during cellular respiration?**

   A ATP

   B CO_2

   C H_2O

   D O_2

2. **What does DNA (deoxyribonucleic acid) do?**

    A DNA controls the structure and function of cells.

    B DNA breaks down food in cells.

    C DNA acts as a stimulus in the environment.

    D DNA acts as a preservative in foods.

3. **Most of the chemical reactions involved in metabolism require**

    A air.

    B oxygen.

    C water.

    D carbon dioxide.

4. **Which of the following are two lipids that organisms use to store energy?**

    A fats and oils

    B phospholipids and oils

    C starches and oils

    D starches and fats

5. **Which of the following is *not* a biochemical?**

    A carbohydrate

    B lipid

    C protein

    D carbon dioxide

SC8.7.a Students know how to identify regions corresponding to metals, nonmetals, and inert gases.

STANDARD REVIEW

In the periodic table, elements are classified as metals, non metals, and metalloids, according to their properties. Regions of the periodic table correspond to classes of elements. The number of electrons in the outer energy level of an atom is one characteristic that helps determine which category an element belongs in.

Most elements are metals. Metals are found to the left of the zigzag line on the periodic table. Atoms of most metals have few electrons in their outer energy level. Most metals are solid at room temperature. Mercury, however, is a liquid at room temperature. Nonmetals are found to the right of the zigzag line on the periodic table. Atoms of most nonmetals have an almost complete set of electrons in their outer level. More than half of the nonmetals are gases at room temperature. Metalloids, also called semimetals, are the elements that border the zigzag line on the periodic table. Atoms of metalloids have about half of a complete set of electrons in their outer energy level.

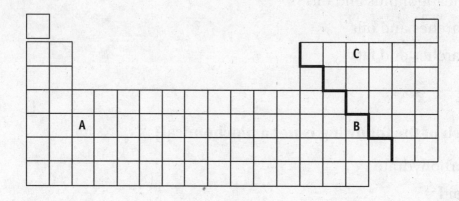

STANDARD PRACTICE

Directions Using the Standard Review and what you have studied, read each question and circle the letter of the best response.

1. **The periodic table on the left shows the placement of three elements being studied in a laboratory. Based on its position on the periodic table, what conclusions can you draw about the likely properties of the element labeled *A*?**

    A The element is a poor conductor and has a shiny appearance.

    B The element is a poor conductor and has a dull looking appearance.

    C The element is a good conductor and has a shiny appearance.

    D The element is a good conductor and has a brittle, dull looking appearance.

2. **The element sulfur (S) has an atomic number of 16, an atomic mass of 32.1, and is a poor conductor of electricity. Based on this information, to which class of elements does sulfur most likely belong?**

    A metals

    B metalloids

    C nonmetals

    D solids

3. **Which of the following elements do *not* usually react with other elements?**

    A carbon group

    B noble gases

    C halogens

    D alkali metals

4. **Which of the following is *not* true of alkali metals?**

    A They can be cut with a knife.

    B They are usually stored in water.

    C They are the most reactive of all the metals.

    D They can easily give away their outer electron.

California Science
Standard 8.7.b
Grade 8

 SC8.7.b Students know each element has a specific number of protons in the nucleus (the atomic number) and each isotope of the element has a different but specific number of neutrons in the nucleus.

STANDARD REVIEW

All of the elements follow the periodic law. The periodic law states that the repeating chemical and physical properties of elements change periodically with the elements' atomic numbers. The atomic number of an element is the number of protons in the nucleus of an atom of that element. All atoms of a given element have the same number of protons in the nucleus. Atoms with different atomic numbers are atoms of different elements. The atomic number is above the chemical symbol of each element on the periodic table. Although each element has a specific number of protons, the number of neutrons for an element can vary. Atoms that have the same number of protons but different numbers of neutrons are isotopes of each other. Each isotope of an element has a specific number of neutrons in the nucleus.

Nuclear Makeup of Different Isotopes

Isotope	Number of Protons	Number of Neutrons
Helium-3	1	2
Beryllium-10	4	6
Radium-226	88	138

STANDARD PRACTICE

Directions Using the Standard Review and what you have studied, read each question and circle the letter of the best response.

1. In the table above, Nuclear Makeup of Different Isotopes, which of the isotopes listed can release energy through the process of beta decay?

    A helium-3

    B helium-3 and beryllium-10

    C beryllium-10 and radium-226

    D radium-226 and helium-3

2. Two atoms of the same element that have different numbers of neutrons in their nuclei are called

    A ions.

    B isotopes.

    C protons.

    D atoms.

3. An atom that has 12 protons, 14 neutrons, and 12 electrons has an atomic number of

    A 12

    B 14

    C 26

    D 38

4. The isotope of uranium used in nuclear reactors, uranium-235, contains 92 protons. Another isotope, uranium-238, contains how many electrons?

    A 92

    B 95

    C 143

    D 146

SC8.7.c Students know substances can be classified by their properties, including their melting temperature, density, hardness, and thermal and electrical conductivity.

STANDARD REVIEW

You probably know a family with several members who look a lot alike. The elements in a family or group in the periodic table often—but not always—have similar properties. The properties are similar because the atoms of the elements in a group have the same number of electrons in their outer energy level. Atoms will often take, give, or share electrons with other atoms in order to have a complete set of electrons in their outer energy level. Elements whose atoms undergo such processes are called reactive. They can combine to form compounds.

Mineral	Magnetite	Quartz	Sulfur
Hardness	6	7	2
Color	Black	Transparent or in a range of colors	Lemon yellow to yellowish brown
Streak	Black	Colorless	White
Luster	Metallic	Glassy	Greasy
Uses	A source of iron used to make steel	Used in making glass and electronic equipment	Used in fungicides, industrial chemicals, and rubber

STANDARD PRACTICE

Directions Using the Standard Review and what you have studied, read each question and circle the letter of the best response.

1. **The chart above shows the results of a laboratory experiment in which a scientist identified some of the properties and uses of 3 minerals. How did she conduct the experiment that determined the streak of each mineral?**

    A She immersed the mineral in water.

    B She broke the mineral along cleavage lines.

    C She rubbed the mineral against a hard plate.

    D She scratched the mineral with a harder mineral.

2. **Which of the following statements is true?**

   A Alkaline-earth metals are more reactive than alkali metals.

   B Alkaline-earth metals have greater density than alkali metals.

   C Alkaline-earth metals have lower atomic numbers than alkali metals.

   D Alkaline-earth metals are more explosive than alkali metals.

3. **What is necessary for substances to burn?**

   A hydrogen

   B oxygen

   C helium

   D carbon

4. **Which one of the following tells the physical state of an element at room temperature?**

   A the atomic number

   B the color of the chemical symbol

   C the atomic mass

   D the element name

5. **Diamond and soot are very different, yet both are natural forms of**

   A carbon.

   B boron.

   C nickel.

   D copper.

California Science
Standard 8.8.a

Grade 8

 SC8.8.a Students know density is mass per unit volume.

STANDARD REVIEW

Which would you rather have fall on your foot—a brick or a foam block that is the size of the brick? Even though both are the same size, the mass of the brick is much greater than the mass of the foam block! You can identify each block by its density. Density is the mass of an object per unit volume. Density equals mass divided by volume. Because 1.0 g of water has a volume of 1.0 mL , water has a density of 1 g/mL.

Densities of Common Substances*			
Substance	**Density (g/cm³)**	**Substance**	**Density (g/cm³)**
Water (liquid)	1.00	Silver (solid)	10.50
Iron pyrite (solid)	5.02	Lead (solid)	11.35
Zinc (solid)	7.13	Mercury (liquid)	13.55
Copper (solid)	8.96	Gold (solid)	19.32
*at 20°C and normal atmospheric pressure			

STANDARD PRACTICE

Direction Using the Standard Review and what you have studied, read each question and circle the letter of the best response.

1. **The table shown above was made during a laboratory investigation on density. What is the difference in density between gold and silver in g/cm³?**

   A 3.85 g/cm³

   B 5.77 g/cm³

   C 8.82 g/cm³

   D 9.22 g/cm³

2. **How do you find out how dense something is?**

   A divide mass by volume

   B divide volume by mass

   C multiply volume by mass

   D multiply mass by volume

3. **Why don't most substances float in air?**

   A Air is too dense.

   B Most substances are denser than air.

   C Air weighs too much.

   D Volume is too great.

4. **If you change the amount of a substance, what happens to its density?**

   A Its density may change.

   B It will get denser.

   C It will get less dense.

   D Its density will not change.

5. **Which of the following statements about density is true?**

   A Density is expressed in grams.

   B Density is mass per unit volume.

   C Density is expressed in milliliters.

   D Density is a chemical property.

SC8.8.b Students know how to calculate the density of substances (regular and irregular solids and liquids) from measurements of mass and volume.

STANDARD REVIEW

Density is a physical property that describes the relationship between mass and volume. Density is the amount of matter in a given space or volume. A golf ball and a table-tennis ball have similar volumes. But a golf ball has more mass than a table-tennis ball does. So, the golf ball has a greater density than the table-tennis ball does. To find an object's density (D), first measure its mass (m) and volume (V). Then, use the equation

$$density = mass/volume$$

Units for density consist of a mass unit divided by a volume unit. The density units most often used are grams per cubic centimeter (g/cm³) for solids and grams per milliliter (g/mL) for liquids. The density of a given substance remains the same no matter how much of the substance you have. That is, the density of 1 cm³ of a substance is equal to the density of 1 km³ of that substance.

MASS AND VOLUME OF VARIOUS SPORTS BALLS

	Mass in Grams (g)	Volume in Cubic Centimeters (cm³)
Golf Ball	45	41
Tennis Ball	57	69
Baseball	145	209
Soccer Ball	400	14 827

STANDARD PRACTICE

Directions Using the Standard Review and what you have studied, read each question and circle the letter of the best response.

1. The table above provides information on the mass and volume of four different types of balls. Which ball has the greatest density?

   A golf ball

   B tennis ball

   C baseball

   D soccer ball

2. Josie determined the mass of 5 cm³ of each of the following substances and got these values:

   Aluminum 13.5 g

   Diamond 17.5 g

   Water 5.0 g

   Wax 4.5 g

   Which substance has the highest density?

   A aluminum

   B diamond

   C water

   D wax

3. **How much more dense is water than air?**

   A 10 times   **C** 1,000 times

   B 100 times   **D** 8,000 times

4. **Which procedure allows you to determine the volume of an irregularly shaped object?**

   A Divide the density of the object by the mass of the object.

   B Calculate the water pressure around the object.

   C Measure the amount of water displaced by the object.

   D Multiply the width, length, and height of the object.

SC8.8.c Students know the buoyant force on an object in a fluid is an upward force equal to the weight of the fluid the object has displaced.

STANDARD REVIEW

Suppose a rock weighs 75 N. It displaces 5 L of water. Archimedes' principle states that the buoyant force is equal to the weight of the displaced water—about 50 N. The rock's weight is greater than the buoyant force. So, the rock sinks. Suppose a fish weighs 12 N. It displaces a volume of water that weighs 12 N. Because the fish's weight is equal to the buoyant force, the fish floats. Now, look at a duck. The duck weighs 9 N. The duck floats. So, the buoyant force on the duck must equal 9 N. But only part of the duck has to be below the surface to displace 9 N of water. So, the duck floats on the surface of the water. If it dives underwater, the duck will displace more than 9 N of water. So, the buoyant force on the duck will be greater than the duck's weight. When the buoyant force on the duck is greater than the duck's weight, the duck is buoyed up (pushed up). An object is buoyed up until the part of the object underwater displaces an amount of water that equals the object's entire weight. Thus, an ice cube pops to the surface when it is pushed to the bottom of a glass of water.

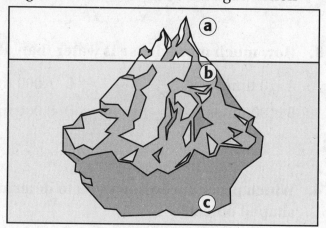

STANDARD PRACTICE

Directions Using the Standard Review and what you have studied, read each question and circle the letter of the best response.

1. **How much of the iceberg in the drawing above has a weight that is equal to the buoyant force?**

    A all of it

    B the section from *a* to *b*

    C the section from *b* to *c*

    D none of the above

2. **What does it mean when a rock sinks in water?**

   A Buoyant force is less than the rock's weight.

   B Buoyant force equals the rock's weight.

   C Buoyant force is greater than the rock's weight.

   D Buoyant force and the rock's weight are unrelated.

3. **What does it mean when a fish floats in water?**

   A Buoyant force is less than the fish's weight.

   B Buoyant force equals the fish's weight.

   C Buoyant force is greater than the fish's weight.

   D Buoyant force and the fish's weight are unrelated.

4. **What does it mean when a duck is buoyed up in water?**

   A Buoyant force is less than the duck's weight.

   B Buoyant force equals the duck's weight.

   C Buoyant force is greater than the duck's weight.

   D Buoyant force and the duck's weight are unrelated.

5. **An object weighs 50 N. When the object is placed in water, the object displaces a volume of water that weighs 10 N. What is the buoyant force on the object?**

   A 60 N

   B 50 N

   C 40 N

   D 10 N

SC8.8.d Students know how to predict whether an object will float or sink.

STANDARD REVIEW

Archimedes was a Greek mathematician who lived in the third century BCE. He discovered how to find buoyant force. Archimedes' principle states that the buoyant force on an object in a fluid is an upward force equal to the weight of the fluid that the object takes the place of, or displaces. Suppose a block displaces 250 mL of water. The weight of 250 mL of water is about 2.5 N. The weight of the displaced water is equal to the buoyant force acting on the block. So, the buoyant force on the block is 2.5 N. Notice that you need to know only the weight of the water that is displaced to find the buoyant force. You do not need to know the weight of the block. But in order to predict if an object will float or sink, you need to consider the weights of both the displaced water and the object.

An object in a fluid will sink if the object's weight is greater than the buoyant force (the weight of the fluid that the object displaces). An object floats only when the buoyant force on the object is equal to the object's weight.

STANDARD PRACTICE

Directions Using the Standard Review and what you have studied, read each question and circle the letter of the best response.

1. **What is it about a ship that allows it to float?**

   A its passengers   **C** its hull

   B its shape   **D** its material

2. **Look at the drawing of the ship on the left. Which of the following statements about the picture is *not* true?**

   A The buoyant force acting on the boat is equal to the force of gravity acting on the boat.

   B The fluid pressure beneath the boat is greater than the fluid pressure on top of the boat.

   C The overall density of the boat is less than the overall density of the water.

   D The boat displaces a volume of water that weighs less than the weight of the boat.

3. **What will happen if something is not as dense as water?**

   A It will sink.   **C** It will float.

   B It will foam.   **D** It will change color.

4. **Which physical property can you use to determine if a substance will float in water?**

   A density   **C** malleability

   B volume   **D** conductivity

5. **Water is an unusual compound because it exists naturally on Earth in three states: solid, liquid, and gas. In which of the following are the water molecules farthest apart?**

   A in an iceberg   **C** in river water

   B in a raindrop   **D** in vapor in the air

California Science
Investigation and Experimentation Standard 8.9   Grade 8

For more practice with the Investigation and Experimentation standards, please refer to the Science Skills Activities in your Student Edition.

 SC8.9 Scientific progress is made by asking meaningful questions and conducting careful investigations. As a basis for understanding this concept and addressing the content in the other three strands, students should develop their own questions and perform investigations.

STANDARD PRACTICE

Directions Read each question and circle the letter of the best response.

1. **What is the first step of a scientific investigation?**

   A conducting an experiment   **C** drawing a conclusion

   B asking a question   **D** making a prediction

   [8.9.a]

2. **A scientist conducts an experiment to test a hypothesis. During the experiment, the scientist takes careful notes of the procedures followed and the results obtained. After the scientist completes the experiment, she shares this information with others who work at the same laboratory. Why is it important for a scientist to keep accurate records of the procedure and results of an experiment?**

   A Accurate records explain how the hypothesis was formed.

   B Accurate records show why a hypothesis is not supported.

   C Accurate records allow others to repeat the experiment.

   D The records help other scientists save time by not repeating the experiment.

   [8.9.b]

3. **When scientists cannot use a controlled experiment to test something, what do they depend on?**

   A variables   **C** observations

   B prediction   **D** hypothesis

   [8.9.c]

4. **Joaquin wants to learn about the effect different amounts of sunlight have on plants. He sets up tables with eight different plants. Every part of his experiment should be the same except the variable he is testing. What is that variable?**

   A the type of plant

   B the amount of light each plant may have

   C the amount of water received by each plant

   D the amount of nutrients he gives each plant

   [8.9.c]

5. **The graph below shows the change in the population of Appleton, Wisconsin between 1900 and 2000. The slope is the ratio of the change in y divided by the change in x. What was the change in the population in Appleton?**

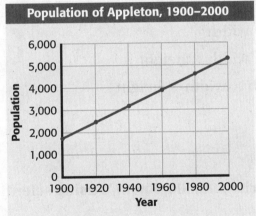

Population of Appleton, 1900–2000

   A The population increased by a fixed amount each year from 1900 to 2000.

   B The population increased, then decreased between 1900 and 2000.

   C The population remained constant between 1900 and 2000.

   D The population decreased, then increased between 1900 and 2000.

   [8.9.d]

6. Jacquelyn conducted an experiment to find out what happens to the temperature of water after it starts boiling. She measured the temperature of the water every 30 seconds during the experiment. She wants to graph her data. If she puts temperature on the y-axis, what variable belongs on the x-axis?

   A water

   B state of matter

   C time

   D boiling point of water

   [8.9.e]

7. What equation would be used to calculate the amount of wallpaper you would need for your classroom?

   A area = length + width

   B area = length × width

   C area = length × width × height

   D volume = length × width × height

   [8.9.f]

8. How would you write 300,000,000 m/s using scientific notation?

   A 3.0×10^8 m/s

   B 30×10^8 m/s

   C $3.0 \times 10 + {}^-8$ m/s

   D $3 \times 100,000,000$ m/s

   [8.9.f]

9. The graph below shows a computer model of human population
    growth from 1000 to 2000 CE. Which pattern is supported by the graph?

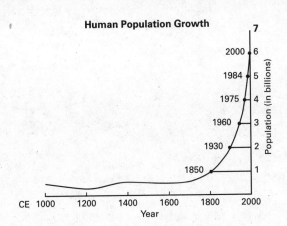

Human Population Growth

A The human population declined steadily over the time shown.

B Population grew most between 1000 and 1600 CE.

C Population size remained constant between 1000 and 1600 CE and then
grew rapidly.

D Between 1850 and 2000 CE, the population increased by 4 billion
people.

[8.9.g]

California Science
Post Standards Assessment

1. **Acceleration has a particular meaning in science. What is the meaning of acceleration as it is used in science?**

   A an object accelerates if its speed increases

   B an object accelerates if its speed increases or decreases

   C an object accelerates if its speed and/or direction change

   D an object accelerates if its speed increases or its direction changes

2. **Dmitri Mendeelev developed the first periodic table of the elements. How did the arrangement of elements in the table devised by Mendeelev differ from the arrangement used today?**

   A Mendeelev arranged elements according to increasing atomic number and today's table organizes elements according to increasing atomic mass.

   B Mendeelev arranged elements according to increasing atomic mass and today's table organizes elements according to increasing atomic number.

   C Mendeelev arranged elements according to their number of neutrons and today's table organizes elements according to their number of protons.

   D Mendeelev arranged elements according to their number of neutrons and today's table organizes elements according to their number of electrons.

3. **The three types of joints involved in walking are**

   A the pivot joint, hinge joint, and glide joint.

   B the ball-and-socket joint, hinge joints, and glide joint.

   C the ball-and-socket joint, hinge joint, and pivot joint.

   D the glide joint, hinge joint, and pivot joint.

4. When an object changes position over time when compared with a reference point, the object is

A accelerating.

B in motion.

C stopping.

D turning.

5. If a man has a malfunction in the cones of the eye that make him unable to distinguish the colors red and green, he has

A a color deficiency.

B nearsightedness.

C eyes that are too short.

D a defective eye lens.

6. Which of the following statements about an element that borders the zigzag line on the periodic table is false?

A The element is likely to be a metalloid.

B The element is likely to have both metallic and nonmetallic properties.

C The element is likely to be extremely unreactive.

D The element is likely to be less malleable and more brittle than most metals.

California Science
Post Standards Assessment

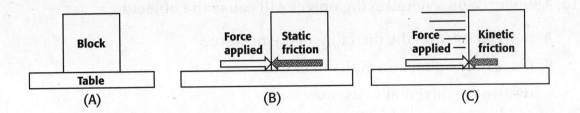

7. **Look at Figure** *A* **above. Why does the block not move?**

    A no force is being applied

    B surface friction

    C force of friction

    D kinetic friction

8. **Look at Figure** *B* **above. The block does not move. What force keeps the block from moving?**

    A rolling kinetic friction

    B static friction

    C sliding kinetic friction

    D kinetic friction

9. **Look at Figure** *C* **above. The block is moving. What force acts against the block's motion?**

    A static friction

    B sliding kinetic friction

    C rolling kinetic friction

    D gravity

10. A net force on a nonmoving object will cause the object to

    A move away from the direction of the net force.

    B move in the direction of the net force.

    C reverse the direction of its movement.

    D reduce the speed of its movement.

11. Look at the drawing below. How many neutrons are found in a carbon-14 nucleus?

Carbon-14 Nucleus

    A 6

    B 8

    C 10

    D 14

12. What type of simple machine is a screwdriver that is used to pry off the lid on a paint can?

    A a pulley

    B a wheel

    C a lever

    D a screw

13. **During an experiment, Mimi observed that a substance is a solid at room temperature, has a high melting point, and dissolves easily in water. Which of the following is a valid conclusion about this substance?**

    A The substance is an acid.

    B The substance is an ionic compound.

    C The substance is a base.

    D The substance is a covalent compound.

14. **Potassium iodide is soluble in water, and has a high melting point of 680°C. The atoms or ions in this compound are held together by**

    A sticky carbon-chained molecules.

    B gravitational forces between the nuclei.

    C the bonds between ions of opposite charges.

    D the attraction between protons and neutrons.

15. **How far away is the most distant object we can see?**

    A about 1 billion light-years

    B 9.46 billion light-years

    C more than 10 billion light-years

    D about 100 billion light-years

16. **Whether an object is transparent, translucent, or opaque is determined by its ability to**

    A refract light.

    B diffract light.

    C transmit light.

    D increase the speed of light.

17. **What are the three major types of galaxies identified by Edwin Hubble?**

    A spiral, triangular, irregular

    B spiral, elliptical, irregular

    C spiral, triangular, elliptical

    D triangular, elliptical, irregular

18. **Which of the following best represents acceleration as presented on a graph?**

    A motion change vs. time

    B distance change vs. time

    C speed change vs. time

    D velocity change vs. time

19. Refraction can separate white light into

    A infrared waves.

    B different colors.

    C ultraviolet light.

    D longer wavelengths.

20. Which of the following would not affect the level at which a canoe floats in a body of water?

    A depth of the water

    B number of people in the canoe

    C shape of the canoe

    D density of the canoe's material

21. Which of the following correctly shows the order of the inner planets according to increasing distance from the sun?

    A Mercury, Venus, Earth, and Mars

    B Mercury, Mars, Venus, and Earth

    C Mercury, Earth, Mars, and Venus

    D Mars, Venus, Earth, and Mercury

22. The reason mass and weight are the same everywhere on Earth is because

    A mass and weight are the same thing.

    B mass and weight are measured the same way.

    C the gravitational force is the same.

    D mass is the same as weight throughout the universe.

23. Which process causes substances to react to form one or more new substances?

    A chemical change

    B physical change

    C evaporation

    D freezing

24. When light reflects off an object, your brain sees the light as traveling

    A in a straight line.

    B in a wavy line.

    C in an S-shaped line.

    D in a curved line.

California Science
Post Standards Assessment

25. Which type of radiation from the sun causes wrinkles and skin cancer?

    A infrared

    B heat

    C ultraviolet

    D visible light

26. A bus is moving north at 15 m/s, and you are walking to the rear of the bus at 1 m/s. Your resultant velocity is

    A 14 m/s north.

    B 16 m/s south.

    C 1 m/s south.

    D 15 m/s north.

27. Ball B is moving to the west. If a force toward the west acts on ball B

    A the ball would have negative acceleration.

    B the velocity of the ball would decrease.

    C the graph of the ball's motion would not change.

    D the ball would travel a greater distance every second.

28. **The figure below shows the Hertzsprung-Russell Diagram. According to the diagram, which of the following is a star that is yellow and has an average brightness?**

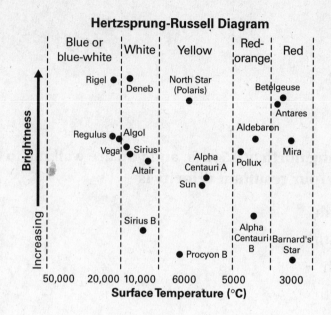

Hertzsprung-Russell Diagram

A Procyon B

B the sun

C Polaris

D Antares

29. **An object moving in a straight line with a constant speed has no unbalanced forces acting on it. How will the object's motion change over time?**

A The object will gradually slow down and come to a stop.

B Centripetal force will cause the object to go into a circular orbit.

C The object's motion will remain unchanged.

D The object will move in a direction opposite the applied force.

30. Which of the following is an exothermic change?

A ice melting

B ice freezing

C dry ice becoming a gas

D water evaporating

31. Aluminum-27 has 13 protons. What is the total mass of its neutrons?

A 13 amu

B 27 amu

C 14 amu

D 40 amu

32. When sand is heated to high temperatures and forms molten glass, it is an example of

A a physical change.

B a chemical change.

C a material change.

D a physical reaction.

33. Balancing a chemical equation so that the same number of atoms of each element is found in both the reactants and the products is an example of

   A activation energy.

   B the law of conservation of energy.

   C the law of conservation of mass.

   D a double-displacement reaction.

34. The vessel in the heart that carries deoxygenated blood is

   A the pulmonary artery.

   B a pulmonary vein.

   C part of the aorta.

   D part of the atria.

35. In a saturated hydrocarbon, how many atoms does each carbon atom bond with?

   A two

   B three

   C four

   D one

36. What is the density of a liquid that has a mass of 206 g and a volume of 321 mL?

    A 1.65 g/mL

    B 1.56 g/mL

    C 0.64 g/mL

    D 0.46 g/mL

37. Scientists often use what types of calculations to determine what kind of substance they are dealing with?

    A mass

    B density

    C weight

    D length

38. Buoyant force is an upward force that fluids exert on matter. Gravity is a downward force that is exerted on all matter. When an object is placed in water, both buoyant force and gravity act on the object. What must be true if an object sinks in water?

    A Buoyant force is equal to gravity.

    B Buoyant force is greater than gravity.

    C Buoyant force is less than gravity.

    D Buoyant force does not act against gravity.

39. Why are most transition metals listed in the chart below better than alkali metals for making jewelry?

Transition Metals

Element	Uses
Iron (Fe)	Manufacturing, building materials, dietary supplements
Cobalt (Co)	Magnets, heat-resistant tools
Nickel (Ni)	Coins, batteries, jewelry, plating
Copper (Cu)	Electric wiring, plumbing, motors
Silver (Ag)	Jewelry, dental fillings, mirror backing, electricity conductors
Gold (Au)	Jewelry, base for money systems, coins, dentistry

A Transition metals are gases at room temperatures.

B Some transition metals produce a magnetic field.

C Transition metals are less reactive than alkali metals.

D Transition metals are silver-colored.

40. What instrument breaks a star's light into a spectrum?

A a continuous spectrum

B a prism

C a spectrometer

D a spectrograph

California Science
Post Standards Assessment

41. What happens when the lens focuses light behind the retina?

   A nearsightedness

   B farsightedness

   C blindness

   D normal vision

42. The state of matter that has a definite volume but no definite shape is

   A solid.

   B liquid.

   C gas.

   D plasma.

43. What is it called when a beam of light arrives at a surface?

   A reflection

   B refraction

   C incidence

   D transmission

44. Which of the following is not a property of a base?

   A It tastes sour.

   B It feels slippery.

   C It changes the color of red litmus paper to blue.

   D It will conduct electricity in a solution.

45. Baptiste walked 420 m in 17.5 min. What is his average speed in m/min? What is his average speed in m/s? (Hint: There are 60 s in 1 min.)

A 18 m/min   0.30 m/s

B 15 m/min   0.25 m/s

C 24 m/min   0.40 m/s

D 45 m/min   0.75 m/s

46. What type of biochemical does not dissolve in water?

A protein

B lipid

C carbohydrate

D vitamin

47. A 5 kg object has less inertia than an object that weighs

A 4 kg.

B 6,000 g.

C 2 kg.

D 1,500 g.

48. Carbon has a central role in the chemistry of living organisms due to the fact that

A carbon helps organisms to regulate temperature.

B carbon atoms allow organisms to transmit nerve impulses.

C carbon is the most abundant element in organisms.

D carbon atoms combine in many ways with atoms of carbon and other elements.

49. When enough energy is added to a liquid that the motion of the particles overcomes the attractions between the particles, the liquid changes into a

A solid.

B plasma.

C gas.

D crystal.

50. What is the distance traveled divided by the time interval during which the motion occurs?

A motion

B reference point

C duration

D speed